Arne-Christian Voigt

Heteronuclear Molecules from a Quantum Degenerate Fermi-Fermi Mixture

Arne-Christian Voigt

Heteronuclear Molecules from a Quantum Degenerate Fermi-Fermi Mixture

A Mass-Imbalanced Fermi System

Südwestdeutscher Verlag für Hochschulschriften

Impressum/Imprint (nur für Deutschland/ only for Germany)
Bibliografische Information der Deutschen Nationalbibliothek: Die Deutsche Nationalbibliothek verzeichnet diese Publikation in der Deutschen Nationalbibliografie; detaillierte bibliografische Daten sind im Internet über http://dnb.d-nb.de abrufbar.

Alle in diesem Buch genannten Marken und Produktnamen unterliegen warenzeichen-, marken- oder patentrechtlichem Schutz bzw. sind Warenzeichen oder eingetragene Warenzeichen der jeweiligen Inhaber. Die Wiedergabe von Marken, Produktnamen, Gebrauchsnamen, Handelsnamen, Warenbezeichnungen u.s.w. in diesem Werk berechtigt auch ohne besondere Kennzeichnung nicht zu der Annahme, dass solche Namen im Sinne der Warenzeichen- und Markenschutzgesetzgebung als frei zu betrachten wären und daher von jedermann benutzt werden dürften.

Verlag: Südwestdeutscher Verlag für Hochschulschriften Aktiengesellschaft & Co. KG
Dudweiler Landstr. 99, 66123 Saarbrücken, Deutschland
Telefon +49 681 37 20 271-1, Telefax +49 681 37 20 271-0
Email: info@svh-verlag.de
Zugl.: München, Ludwig-Maximilians-Universität München, Dissertation, 2009

Herstellung in Deutschland:
Schaltungsdienst Lange o.H.G., Berlin
Books on Demand GmbH, Norderstedt
Reha GmbH, Saarbrücken
Amazon Distribution GmbH, Leipzig
ISBN: 978-3-8381-1273-2

Imprint (only for USA, GB)
Bibliographic information published by the Deutsche Nationalbibliothek: The Deutsche Nationalbibliothek lists this publication in the Deutsche Nationalbibliografie; detailed bibliographic data are available in the Internet at http://dnb.d-nb.de.

Any brand names and product names mentioned in this book are subject to trademark, brand or patent protection and are trademarks or registered trademarks of their respective holders. The use of brand names, product names, common names, trade names, product descriptions etc. even without a particular marking in this works is in no way to be construed to mean that such names may be regarded as unrestricted in respect of trademark and brand protection legislation and could thus be used by anyone.

Publisher: Südwestdeutscher Verlag für Hochschulschriften Aktiengesellschaft & Co. KG
Dudweiler Landstr. 99, 66123 Saarbrücken, Germany
Phone +49 681 37 20 271-1, Fax +49 681 37 20 271-0
Email: info@svh-verlag.de

Printed in the U.S.A.
Printed in the U.K. by (see last page)
ISBN: 978-3-8381-1273-2

Copyright © 2010 by the author and Südwestdeutscher Verlag für Hochschulschriften Aktiengesellschaft & Co. KG and licensors
All rights reserved. Saarbrücken 2010

Meinen Eltern

Abstract

This thesis reports on experiments with quantum degenerate atomic mixtures and molecular gases at ultracold temperatures. The work describes the first realization of a quantum degenerate Fermi-Fermi mixture of two different species with unequal masses. In addition, the first quantum degenerate three-species mixture is realized. Furthermore, the first heteronuclear bosonic molecules with temperatures close to quantum degeneracy are created from a two-species mixture.

Within this work a new and very versatile experimental platform to study quantum degenerate two-species Fermi mixtures is presented. The experimental concept relies on sympathetic cooling of two fermionic species (^{40}K and ^{6}Li) by a large bosonic gas (^{87}Rb). It is shown that large atom numbers and reliable operation are guaranteed by careful choice of the experimental components and parameters, which is essential to deal with the complexity of such an experimental platform with three atomic species.

The first important milestone towards quantum degeneracy is realized by simultaneous trapping ^{87}Rb, ^{40}K and ^{6}Li in a magneto-optical trap. This marks the first realization of magneto-optical trapping of two fermionic species and also of three species. To achieve quantum degeneracy, different experimental challenges and difficulties of the three species with very different initial temperatures, scattering cross sections and masses were overcome. A combined compressed MOT and temporal dark MOT phase, a careful state cleaning process during the cooling process, species selective evaporative cooling of rubidium and removing of high energetic lithium atoms turn out to be essential for the successful realization of a quantum degenerate Fermi-Fermi-Bose mixture. Furthermore, at the end of the sympathetic cooling process, the cooling efficiency of lithium is improved by catalytic cooling through potassium.

A quantum degenerate Fermi-Fermi mixture can be realized in the strongly interacting regime at a s-wave Feshbach resonance. By an adiabatic magnetic field sweep across the Feshbach resonance ultracold bosonic heteronuclear molecules are created with high efficiencies of up to 50 %. The associated number of molecules are up to 4×10^4 molecules, which is the largest number of heteronuclear molecules produced so far. Furthermore, close to resonance an increased molecular lifetime of more than 100 ms is observed. Therefore, this is the first system that may be used to explore many-body physics of a heteronuclear mixture in the strongly interacting regime across a Feshbach resonance.

The heteronuclear Fermi mixture adds the mass ratio and the different internal structure as new parameters to quantum degenerate Fermi gases. The molecular cloud being close to quantum degeneracy marks the first starting point for the realization of a heteronuclear molecular BEC and to investigate the BEC-BCS crossover with unequal masses. Moreover, the molecules can be transferred into the rovibrational ground state to create a polar BEC with an anisotropic, long-range interaction.

Zusammenfassung

Die vorliegende Arbeit beschreibt Experimente mit quantenentarteten atomaren Mischungen und molekularen Gasen bei ultrakalten Temperaturen. Die Arbeit zeigt die erste Realisierung einer quantenentarteten Fermi-Fermi-Mischung aus zwei verschiedenen Atomsorten als auch die Herstellung einer dreifach quantenentarteten Mischung. Darüber hinaus werden zum ersten Mal heteronukleare bosonische Moleküle, deren Temperatur nahe an der Quantenentartung liegt, aus zwei unterschiedlichen Atomsorten hergestellt.

Im Rahmen dieser Arbeit wird eine neue, vielseitig einsetzbare experimentelle Plattform zum Studium von quantenenarteten fermionischen Mischungen aus unterschiedlichen Atomsorten präsentiert. Das experimentelle Konzept beruht auf der sympathetischen Kühlung von zwei fermionischen Atomsorten (^{40}K und ^{6}Li) mit einem großen bosonischen Gas (^{87}Rb). Es wird gezeigt, daß große Atomzahlen und ein zuverlässiges Arbeiten durch eine sorgfältige Wahl der experimentellen Komponenten und Parameter erfüllt werden kann. Dies ist essenziell, um die Komplexität eines solchen Experiments mit drei Spezies zu bewältigen.

Der erste wichtige Meilenstein zur Quantenentartung ist durch das simultane Fangen von ^{87}Rb, ^{40}K und ^{6}Li in einer magneto-optischen Falle realisiert. Dies stellt das erste magnetooptische Fangen von zwei fermionischen als auch von drei Atomsorten dar. Zur Quantenentartung wurden unterschiedlichste experimentelle Herausforderungen mit den drei Atomsorten, die verschiedene Anfangstemperaturen, Streuquerschnitte und Massen aufweisen, überwunden. Für eine erfolgreiche Herstellung einer quantenentarteten Fermi-Fermi-Bose-Mischung sind eine Kombination aus komprimierter MOT und zeitlicher Dunkel-MOT Phase, eine sorgfältige Zustandsreinigung während der Kühlphase, speziesselektive Verdampfungskühlung von Rubidium und Entfernung von hochenergetischen Lithium Atomen entscheidend. Ferner wird die Kühleffizienz von Lithium durch katalytisches Kühlen durch Kalium verstärkt.

Eine quantenentartete Fermi-Fermi-Mischung kann im starkwechselwirkendem Regime an einer s-Wellen-Feshbach-Resonanz präpariert werden. Mit Hilfe einer adiabatischen Magnetfeldveränderung über die Feshbach-Resonanz können ultrakalte bosonische Moleküle mit hohen Effizienzen bis zu 50 % hergestellt werden. Die assoziierten Molekülzahlen sind bis zu 4×10^4, was die derzeit größten heteronuklearen Molekülzahlen darstellt. Darüberhinaus weisen die Moleküle eine ansteigende Lebensdauer nahe der Resonanz von mehr als 100 ms auf. Demzufolge ist es das erste System, mit dem Vielteilchenphysik einer heteronuklearen Mischung im starkwechselwirkenden Regime an einer Feshbach-Resonanz studiert werden kann.

Die heteronukleare Fermi-Mischung erweitert die quantenenarteten Fermi-Gase um das Massenverhältnis und die unterschiedliche interne Struktur. Das molekulare Gas, welches nahe an der Quantenentartung ist, markiert den ersten Startpunkt hin zu einem heteronuklearen Molekül-BEC und der Untersuchung des BEC-BCS-Übergangs mit unterschiedlichen Massen. Darüberhinaus können die Moleküle in den rovibrationellen Grundzustand überführt werden, um ein polares BEC mit einer anisotropen, langreichweitigen Wechselwirkung herzustellen.

Contents

Abstract		**vii**
1	**Introduction**	**1**
	1.1 Bose-Einstein condensation	1
	1.2 Quantum degenerate Fermi gases	2
	1.3 Many-body physics and the BEC-BCS crossover	4
	1.4 Ultracold chemistry	5
	1.5 This thesis	7
2	**Theory**	**9**
	2.1 Ultracold gases	9
	2.1.1 Quantum statistical properties	9
	2.1.2 Bosons	11
	2.1.3 Fermions	13
	2.2 Feshbach resonances	14
	2.2.1 Ultracold Collisions	14
	2.2.2 Magnetic Feshbach resonance	17
	2.2.3 The asymptotic bound state model	18
	2.2.4 Classification of Feshbach resonances	20
	2.2.5 Fermi-Fermi mixtures	22
3	**Experimental Setup**	**25**
	3.1 Experimental concept	25
	3.2 Vacuum system	26
	3.3 Atomic sources	27
	3.3.1 Dispensers for Rb and K	28
	3.3.2 Zeeman slower	28
	3.4 Laser systems	29
	3.4.1 Energy levels	29
	3.4.2 Rubidium	29
	3.4.3 Potassium	31
	3.4.4 Lithium	32
	3.4.5 Optical system for MOT, detection and optical pumping	33
	3.4.6 Optical dipole trap	33
	3.5 Absorption imaging	36
	3.5.1 Principle	36
	3.5.2 Optical setup	37

		3.5.3	State selective imaging	38
	3.6	Magnetic trapping		38
		3.6.1	Principle	38
		3.6.2	Magnetic transport	39
		3.6.3	QUIC trap	40
	3.7	Feshbach coils		41
	3.8	Ambient magnetic fields		43
		3.8.1	Magnetic offset fields	43
		3.8.2	External magnetic fields	44
	3.9	RF & MW setup		45
	3.10	Experimental Control		47

4 Magneto-optical trapping of three atomic species — 49
- 4.1 Experimental setup — 49
- 4.2 The triple MOT — 50
  - 4.2.1 Single MOTs — 50
  - 4.2.2 Triple MOT — 51
  - 4.2.3 Dispenser currents — 51
  - 4.2.4 Light-assisted collisions — 52
  - 4.2.5 Optical molasses — 53
- 4.3 Further optimization and conclusions — 53

5 Quantum degeneracy — 55
- 5.1 On the road to quantum degeneracy — 55
  - 5.1.1 cMOT and dMOT — 55
  - 5.1.2 State preparation — 57
  - 5.1.3 Magnetic transport — 58
  - 5.1.4 QUIC trap — 59
- 5.2 Cooling into quantum degeneracy — 60
  - 5.2.1 Evaporative cooling — 60
  - 5.2.2 Sympathetic cooling — 64
  - 5.2.3 Quantum degenerate Bose-Fermi-Fermi mixture — 67
- 5.3 Conclusions — 71

6 Ultracold heteronuclear Fermi-Fermi molecules — 73
- 6.1 The optical dipole trap — 73
  - 6.1.1 Loading into the ODT — 73
  - 6.1.2 Characterization of the ODT — 74
- 6.2 Magnetic field calibration — 75
- 6.3 State preparation — 77
- 6.4 Identifying Feshbach resonances — 78
  - 6.4.1 Loss measurement — 79
- 6.5 Heteronuclear Fermi-Fermi Molecules — 80
  - 6.5.1 Adiabatic conversion process — 80
  - 6.5.2 Reconversion process — 81
  - 6.5.3 Direct molecule detection and molecule purification — 82
  - 6.5.4 Lifetime measurement — 84

6.6 Discussion & conclusions	85

7 Conclusions and Outlook 87

A Natural constants and atomic sources 89
A.1 Natural constants 89
A.2 Atomic properties 90

Bibliography 93

Danksagung 113

Contents

Chapter 1

Introduction

Since the first realization of Bose-Einstein condensation in ultracold, dilute gases 14 years ago (Anderson et al., 1995; Bradley et al., 1995; Davis et al., 1995), followed by quantum degeneracy of fermions (DeMarco and Jin, 1999), the research field on ultracold, dilute quantum gases has produced many exciting experiments and has opened new prospects on atomic and many-body physics, and also refocussed the interest on molecular physics.

In the following, an overview on ultracold quantum gases and their recent developments is given, first on bosons in Sec. 1.1 and then on fermions in Sec. 1.2. After this their application to many-body physics and ultracold chemistry are discussed in Sec. 1.3 and Sec. 1.4, respectively. Finally, Sec. 1.5 gives an outline of this thesis.

1.1 Bose-Einstein condensation

Based on a work on photon statistics by Satyendra Nath Bose (Bose, 1924), Albert Einstein predicted Bose-Einstein condensation (BEC) for an ideal Bose gas in the year 1925 (Einstein, 1925). This phenomena of condensation is a consequence of quantum statistics and postulates a macroscopic occupation of the single particle ground state for temperatures below a critical value.

The connection between superflluid liquid ^{4}He and Bose-Einstein condensation was first suggested by Fritz London (London, 1938a,b). However, due to large interparticle interactions in liquid helium, the number of atoms in the single particle ground state is reduced and only a phenomenological description of the superfluid phase is possible. A very successful phenomenological model, given by L. D. Landau, is the two-fluid model (Landau, 1949).

In dilute atomic gases, on the other side, interparticle interactions are typically weak and allow to create pure condensates. For typical atomic densities of $10^{13} - 10^{15}$ cm^{-3} the timescale for the formation of molecules and for liquification is much longer than the timescale needed for thermalization by two-body collisions, which means that the two-body collision rate dominates the three-body collision rate. This allows to achieve Bose-Einstein condensation in a metastable gaseous phase for temperatures far below the critical temperature.

The temperature needed for Bose-Einstein condensation in dilute systems are on the order of hundreds of nanokelvin. This makes it experimentally quite challenging to reach this regime. The first important step towards ultralow temperatures was the proposal by Hänsch and Schawlow (1975) and the following experiments in the 1980s on laser cooling of neutral atomic gases (see Nobel lectures by Chu (1998); Cohen-Tannoudji (1998); Phillips (1998)), their

1. Introduction

magnetic trapping (Migdall et al., 1985) and the realization of a magneto-optical trap (MOT) (Raab et al., 1987). Here, the first results for laser cooling were achieved with alkali atoms that have certain advantages due to their relative simple energy-level structure. Another key cooling technique is evaporative cooling (Hess, 1986; Masuhara et al., 1988), which was developed in experiments on spin polarized atomic hydrogen, before it was applied to alkaline gases in magnetic or optical dipole traps (Dalibard and Cohen-Tannoudji, 1985; Chu et al., 1986). This technique allows to increase phase-space density by several orders of magnitude. Current laser cooling techniques are limited in the final phase-space density and up to now evaporative cooling is the only used technique to achieve the quantum degenerate regime.

The first realization of Bose-Einstein condensation was achieved with the alkali species ^{85}Rb, ^{23}Na and ^{7}Li (Anderson et al., 1995; Davis et al., 1995; Bradley et al., 1995). The list of condensed species is still increasing and counts 14 different atomic species now (Fried et al., 1998; Cornish et al., 2000; Robert et al., 2001; Santos et al., 2001; Modugno et al., 2001; Weber et al., 2002; Takasu et al., 2003; Griesmeier et al., 2005; Fukuhara et al., 2007a,c; Sterr, 2009). Dilute Bose-Einstein condensations are prime candidates to study quantum matter phenomena with precise control over key parameters and offer a model system for many-body physics. This has led to a boom of dilute quantum gases with almost one hundred research groups worldwide. The first experiments on Bose-Einstein condensation were focused on characteristic properties like the interference of matter waves, elementary excitations, the quantization of vortices and many more (see also reviews on this subject by Dalfovo et al. (1999); Leggett (2001); Pethick and Smith (2002); Pitaevskii and Stringari (2003)).

One interesting research direction that is relevant in the context of this thesis is to tune the interparticle interaction by a Feshbach resonance, which is originally a concept from nuclear physics (Fano, 1935, 1961; Feshbach, 1958, 1962). A Feshbach resonance occurs, if the collision energy of two atoms energetically coincides with a molecular bound state. Such a Feshbach resonance can be tuned precisely, for example, by an external magnetic field (Tiesinga et al., 1993) or optically by laser light (Fedichev et al., 1996a; Bohn and Julienne, 1997, 1999). On resonance the scattering length a has a singularity, $a \to \pm\infty$, which in principle allows to vary a to arbitrary values. This tunability makes Feshbach resonances a powerful tool in the field of ultracold gases, where the first Feshbach resonances were observed with bosonic atoms (Inouye et al., 1998; Courteille et al., 1998; Roberts et al., 1998; Vuletić et al., 1999). However, working with bosons close to a Feshbach resonance with large interactions is quite challenging. In the vicinity of a Feshbach resonance three-body losses are enhanced for bosonic atoms and limits the molecular lifetime in a bulk gas significantly (Fedichev et al., 1996b; Inouye et al., 1998; Vuletić et al., 1999; Stenger et al., 1999; Roberts et al., 2000). Nevertheless, in a three-dimensional optical lattice it is possible to create isolated molecules on single lattice sites and to increase the molecular lifetime significantly (Thalhammer et al., 2006; Volz et al., 2006).

1.2 Quantum degenerate Fermi gases

After the successful creation of Bose-Einstein condensation in dilute gases and exciting experiments on coherent, macroscopic many-body states, the dramatic progress in the experimental methods made quantum degeneracy in dilute fermionic gases possible. This achievement allows to investigate differences in quantum statistics between fermionic and bosonic systems. Furthermore, fermions offer richer phenomena in the context of many-body physics.

In contrast to bosons, non-interacting fermions have a smooth crossover in their character-

1.2 Quantum degenerate Fermi gases

istic thermodynamic properties, when they are cooled into quantum degeneracy. The quantum degenerate regime for fermions is generally defined for temperatures well below the Fermi temperature $T_F = E_F/k_B$, here E_F is the Fermi energy. In 1999 the first quantum degenerate Fermi gas was experimentally realized with ^{40}K at JILA (DeMarco and Jin, 1999). The list of degenerate Fermi gases is currently enlarged by ^{6}Li (Truscott et al., 2001; Schreck et al., 2001), metastable ^{3}He* (McNamara et al., 2006) and two earth alkali species ^{171}Yb and ^{173}Yb (Fukuhara et al., 2007b).

From the experimental point of view, cooling fermions into quantum degeneracy is more challenging than in the bosonic case. At ultracold temperatures interparticle scattering is dominated by s-wave collisions. But, due to the Pauli exclusion principle, fermionic atoms of a one-component Fermi gas do not scatter with each other, which makes evaporative cooling of a one-component Fermi gas unfeasible. One strategy to circumvent this problem is to evaporate an incoherent spin mixture of fermionic atoms as was used to realize the first quantum degenerate Fermi gas with ^{40}K (DeMarco and Jin, 1999). Another strategy is by sympathetic cooling (Wineland et al., 1978) a one-component Fermi gas that stay in thermal contact with an actively cooled bath. This technique is currently widely used, because the fermionic atom number is in principle not reduced and high atom numbers with low temperatures are achieved.

So far, successful sympathetic cooling of fermions has been demonstrated with the following atomic combinations: ^{6}Li – ^{7}Li (Truscott et al., 2001; Schreck et al., 2001), ^{6}Li – ^{23}Na (Hadzibabic et al., 2002), ^{6}Li – ^{87}Rb (Silber et al., 2005), ^{40}K – ^{87}Rb (Roati et al., 2002. Köhl et al., 2005; Ospelkaus et al., 2006b; Aubin et al., 2006) , ^{3}He* – ^{4}He* (McNamara et al., 2006) and ^{171}Yb – ^{174}Yb (Fukuhara et al., 2007c).

The first experiments on degenerate Fermi gases were concentrated on one-component systems, representing an ideal Fermi gas. These studies explored, for example, thermodynamic properties of the Fermi gas, its deviations from a classical gas, Pauli blocking of collisions at ultralow temperatures (DeMarco and Jin, 1999; DeMarco et al., 2001) and the Fermi pressure of a trapped gas (Truscott et al., 2001).

In fermionic gases Feshbach resonances between two different hyperfine states of an incoherent mixture allow to tune the interparticle interaction. Several Feshbach resonances were found in ^{6}Li (O'Hara et al., 2002b; Dieckmann et al., 2002; Jochim et al., 2002) and ^{40}K (Loftus et al., 2002). By a magnetic field sweep across the Feshbach resonance weakly bound diatomic molecules were produced (Regal et al., 2003). Surprisingly, these molecules showed long lifetimes in the vicinity of a Feshbach resonance (Cubizolles et al., 2003; Jochim et al., 2003a; Strecker et al., 2003; Regal et al., 2004a), even for large interactions, which is contrary to the case of bosons. The observed lifetimes were up to several seconds at high densities of about 10^{13} cm^{-3}.

This stability is a consequence of the Pauli exclusion principle (Petrov et al., 2004), that suppresses vibrational quenching to energetically lower vibrational molecular states. The increased lifetime at resonance led to a boom of Fermi gases and allows to work in the strongly interacting regime $k_F|a| \geq 1$, where the scattering length a is larger or comparable to the interparticle spacing $\sim 1/k_F$ (k_F is the Fermi momentum). This discovery opened the way to explore the crossover regime between Bose-Einstein condensation (BEC) and Bardeen-Cooper-Schrieffer (BCS) superfluidity (Bardeen et al., 1957), which is discussed in the following Sec. 1.3.

1.3 Many-body physics and the BEC-BCS crossover

To study many-body physics in dilute atomic gases the role of interactions is an important issue. Periodic optical potentials and Feshbach resonances, as previously mentioned in Sec. 1.1, are essential tools to access the physics of strongly interacting systems with atomic gases (see also review article on many-body physics by Bloch et al. (2008)).

The ability to shape the dimensionality and to localize atoms with optical potentials is one possibility to enter the regime of strong correlations. An outstanding example is the quantum phase transition from a superfluid to a Mott insulator with a BEC in a three-dimensional optical lattice (Greiner et al., 2002). Other examples with ultracold quantum gases in reduced dimensionality are the realization of a Tonks-Girardeau gas (Kinoshita et al., 2004; Paredes et al., 2004) in one-dimension and the Berezinskii-Kosterlitz-Thouless crossover (Hadzibabic et al., 2006; Schweikhard et al., 2007) in two-dimensions.

Moreover, bosons and fermions in periodic potentials (Modugno et al., 2003) provide the link to condensed matter physics by the (Bose-)Hubbard Hamiltonian. For example, the Fermi surface and the band insulator with fermions are demonstrated (Köhl et al., 2005) and a Mott insulator of fermions is recently realized (Jördens et al., 2008; Schneider et al., 2008). The precise control of the optical potentials offer novel directions in condensed matter physics.

The other possibility to achieve strongly interacting systems are Feshbach resonances that allow to control the interaction by tuning the scattering length a. Long lifetime at the resonance in the bulk gas is a necessary precondition for experimental studies, and only is exhibited for the case of fermions, where decay is suppressed due to Pauli exclusion principle (see discussion in Sec. 1.2). The lifetime can even exceed the thermalization time by collisions. This discovery allowed to create a molecular BEC (mBEC) (Jochim et al., 2003b; Greiner et al., 2003; Zwierlein et al., 2003) of weakly bound molecules composed of fermions on the repulsive, molecular side of the Feshbach resonance ($a > 0$).

Furthermore, the long lifetime at large values of the scattering length a is important to realize BCS pairing on the attractive side of the Feshbach resonance ($a < 0$). In the weak coupling regime $k_\mathrm{F} |a| \ll 1$ the critical temperature T_C scales with a as $T_\mathrm{C} \approx 0.28\, T_\mathrm{F} \exp\left(-\pi/2\, k_\mathrm{F} |a|\right)$ (Gorkov and Melik-Barkhudarov, 1961). Close to resonance, in the strong coupling regime $k_\mathrm{F} |a| \geq 1$, no analytic solution of the many-body problem is available and numerical calculations are required (Nozières and Schmitt-Rink, 1985; Haussmann et al., 2007). In this regime T_C is of order T_F and is thus experimentally accessible. This allowed to observe condensation of fermionic pairs on the BCS-side (Regal et al., 2004b; Zwierlein et al., 2004). Moreover, the tunability of the scattering length allows to study the BEC-BCS crossover (Bartenstein et al., 2004b; Bourdel et al., 2004), going from pairing in real space to pairing in momentum space. This crossover connects the two superfluid regimes smoothly across the strongly interacting regime.

At resonance, the scattering length diverges and the gas is in the so-called unitarity regime. Here, the only energy and length scale of the system are the Fermi energy E_F and Fermi momentum k_F and all thermodynamic quantities are related to universal properties (Baker, 1999; Heiselberg, 2001; Ho, 2004).

Dynamic experiments at the BEC-BCS crossover include the anisotropic expansion of the gas (O'Hara et al., 2002a) and collective excitation measurements (Kinast et al., 2004; Bartenstein et al., 2004a) with aiming to proof superfluidity. However, similar behavior can be experimentally shown in a classical hydrodynamic gas (Clancy et al., 2007; Wright et al., 2007). A first signature of superfluid behavior along the crossover was the observation of the pairing gap by radio-frequency spectroscopy (Chin et al., 2004). The smoking gun experiment for superfluidity was the observation of quantized vortices across the BEC-BCS crossover (Zwierlein et al., 2005).

Currently, experiments concentrate on systems with imbalanced spin-mixtures, that have an unequal occupation of the two spin states. This leads to an unequal chemical potential of the two spin components and exotic quantum phases with different pairing mechanisms are expected in this situation (Fulde and Ferrell, 1964; Larkin and Ovchinnikov, 1965; Sarma, 1963; Liu and Wilczek, 2003; Bedaque et al., 2003; Caldas, 2004; Carlson and Reddy, 2005). In imbalanced mixtures, so far, phase separation is observed (Partridge et al., 2006), fermionic superfluidity with imbalanced spin populations is proven (Zwierlein et al., 2006) and the superfluid phase transition is directly observed (Schunck et al., 2007).

1.4 Ultracold chemistry

Molecules at ultracold temperatures open new perspectives on high precision tests of fundamental physics, on quantum chemistry and on novel quantum matter states (see also review article on ultracold molecules by Carr et al. (2009)). Molecules provide additional degrees of freedom by the rotational and vibrational states, which offer new possibilities in their experimental control. For the first time, ultracold molecules have the potential to prepare a single quantum state of the internal and external degrees of freedom and offer unique ways of high resolution quantum control (Pe'er et al., 2007; Shapiro et al., 2008).

Moreover, dense and ultracold ensembles will allow precise control of chemical reactions and the investigation of ultracold chemistry (Krems, 2005; Hudson et al., 2006b; Krems, 2008). At high phase space densities with large de Broglie wavelengths reaction dynamics are influenced by quantum effects and tunneling processes play an important role.

Of particular interest are ultracold polar molecules with the dipole moment as an extra degree of freedom. In combination with the precise controllability polar molecules can be used to test fundamental physics, for example to study time variation of fundamental constants (Hudson et al., 2006a; Chin et al., 2009) and violation effects of parity and time-reversal (Kozlov and de Mille, 2002; Hudson et al., 2002). Moreover, the dipole-dipole interaction in polar molecular gases is anisotropic and has a long-range character. This opens many new possibilities that range from anisotropically interacting quantum fluids, like a polar molecular BEC, to the exploration of electric dipole-dipole mediated BCS pairing and many more phenomena (Santos et al., 2000, 2002; Góral and Santos, 2002; Góral et al., 2002; Baranov et al., 2002; Damski et al., 2003; Griesmaier et al., 2006). Polar molecules are also interesting for quantum information processing (de Mille, 2002) due to effective coupling of the electric dipoles at moderate distances and the relative easy controllability of the electric dipole moments by electric fields (DC or microwave fields).

The production process of ultracold molecular ensembles is an experimentally challenging task and many different approaches exist. The experimental techniques can be classified into direct or indirect methods. Direct techniques start with stable ground-state molecules that are further cooled in their external, motional degree of freedom. Applying laser cooling methods to molecules would be extremely demanding due to their complex level structure and recent proposals, as (Bahns et al., 1996; Rosa, 2004; Stuhl et al., 2008), are still to be tested. Current strategies of direct cooling include decelerating a molecular beam by external fields, which

1. INTRODUCTION

can be Stark (Bethlem et al., 1999; Bochinski et al., 2003) or Zeeman (Hogan et al., 2007; Narevicius et al., 2008) potentials, or even optical fields (Fulton et al., 2006). Another powerful and versatile strategy is buffer-gas cooling with helium atoms (Weinstein et al., 1998). The achievable temperatures with direct methods are currently in the range of 10 mK and 1 K with relative small phase space densities of up to 10^{-12}. Therefore, to reach ultracold temperatures, further cooling techniques have to be developed, e.g. sympathetic cooling with laser cooled alkalines.

Indirect techniques, on the other hand, start with precooled atoms by cooling methods previously described in Sec. 1.1 and 1.2 (i.e. laser cooling, evaporative and sympathetic cooling). After precooling, the atomic constituents are assembled to molecules by photoassociative (Thorsheim et al., 1987) or Feshbach methods (Mies et al., 2000). Due to technical constraints of trapping and laser cooling, the indirect technique can be used for certain atomic classes, such as alkali and alkaline earth metals. Typically, the molecules that are created with this indirect method are in their energetically highest vibrational state and can perform inelastic relaxation to energetically deeper lying levels (Zirbel et al., 2008a; Hudson et al., 2008), resulting in trap losses and heating. Moreover, the expected electric dipole moments of these heteronuclear molecules are negligible due to their highest vibrational state. The excited molecules can be transferred to the absolute rovibrational ground state (Sage et al., 2005), which is energetically stable against inelastic two-body processes and exhibits an appreciable electric dipole moment in the heteronuclear case. (Note, that the expectation value of the electric dipole moment is only non-zero when a finite electric field is applied.)

The coldest temperatures are currently achieved by Feshbach methods with temperatures down to a few tens of nanokelvin with high phase-space densities and successful realization of BEC with homonuclear molecules (Jochim et al., 2003b; Greiner et al., 2003; Zwierlein et al., 2003). This method marks therefore the most powerful and promising route towards quantum degenerate polar molecules. At a Feshbach resonance weakly bound molecules can be created by three different methods: The first one is by magnetic field sweeps across a Feshbach resonance (van Abeelen and Verhar, 1999; Timmermans et al., 1999; Mies et al., 2000), the second one by oscillatory fields (Thompson et al., 2005; Hanna et al., 2007) and the third one by atom-molecule thermalization (Jochim et al., 2003b; Zwierlein et al., 2003). So far, homonuclear bialkali molecules have been produced of the bosonic species ^{23}Na (Xu et al., 2003), ^{87}Rb (Dürr et al., 2004) and ^{133}Cs (Herbig et al., 2003) and of the fermionic species ^{6}Li (Strecker et al., 2003; Cubizolles et al., 2003) and ^{40}K (Regal et al., 2003). At present, research is concentrated on heteronuclear bialkali molecules. The molecules are currently produced from Bose-Fermi (^{40}K − ^{87}Rb (Ospelkaus et al., 2006a; Zirbel et al., 2008b)) and from Bose-Bose mixtures (^{85}Rb − ^{87}Rb (Papp and Wieman, 2006) and recently ^{41}K − ^{87}Rb (Weber et al., 2008)). These highly excitetd molecules can be transferred to deeply bound states by a coherent stimulated Raman adiabatic passage (STIRAP) (Bergmann et al., 1998), which allows high transfer efficiencies. In 2008, homonuclear ground state molecules of ^{87}Rb$_2$ (Lang et al., 2008) and ^{133}Cs$_2$ (Danzl et al., 2008) were created by this coherent transfer technique. Furthermore, fermionic polar molecules of ^{40}K^{87}Rb in their absolute ground state with phase space densities close to quantum degeneracy were reported in the same year (Ni et al., 2008).

1.5 This thesis

This thesis describes two main achievements that offer new perspectives for future experiments in the context of many-body physics and ultracold chemistry.

Firstly, a quantum degenerate Fermi-Fermi mixture of two different atomic species with unequal masses is realized for the first time. This result adds the mass ratio and the different internal structure as new control parameters to quantum degenerate Fermi gases. Such a Fermi-Fermi mixture allows to study the nature and existence of superfluidity in the presence of unmatched Fermi surfaces, which can be realized by number imbalance or by mass imbalance. In this situation, symmetric BCS pairing is broken and many new quantum phases with different pairing mechanisms are expected, as the Fulde-Ferrell-Larkin-Ovchinnikov (FFLO) state (Fulde and Ferrell, 1964; Larkin and Ovchinnikov, 1965), the breached pair state (Liu and Wilczek, 2003) and the Sarma state (Sarma, 1963). Moreover, a crystalline phase transition (Petrov et al., 2007) and a link to baryonic phases of quantum chromodynamics (Wilczek, 2007; Rapp et al., 2007) are predicted.

Secondly, the first realization of ultracold heteronuclear bosonic molecules of two different fermionic species is shown. This completes the list of possible quantum-statistical combinations in mixtures. The observed long lifetimes of the Fermi-Fermi molecules in the molecule-atom mixture are promising to explore many-body physics with two different masses and, in addition, to create bosonic dipolar molecules in their absolute ground state.

For experimental studies across a Feshbach resonance the molecular lifetime is an important aspect. The stability of molecules composed out of fermions depends on the mass ratio of the two fermions (Petrov et al., 2005). At a Feshbach resonance long molecular lifetimes are expected for mass ratios smaller than 12.33, for which the molecular relaxation decreases with an increasing scattering length. For the experimental studies therefore, the fermionic isotopes of potassium and lithium are chosen (^{40}K and ^{6}Li) that have a mass ratio of 6.67.

To cool the Fermi-Fermi mixture into quantum degeneracy we chose the sympathetic cooling strategy, which uses another atomic cloud as a cooling agent. In our case we decided to use the bosonic alkali atom ^{87}Rb. Sympathetic cooling has the advantage that the atom number of the fermions is in principle not reduced by evaporation, and therefore avoids the need for high number atom sources. Moreover, evaporation of fermions would have the disadvantage of Pauli blocking at the final stage of evaporation. In addition, the expected dipole moment for ground state molecules of LiK and LiRb (Aymar and Dulieu, 2005) is larger than in comparable existing experiments and should allow to study quantum gases with large dipolar interactions.

The following chapters of this thesis are organized as follows:

- Chapter 2 gives an theoretical overview on the quantum statistical and thermodynamic properties of bosons and fermions. Then collisional processes in ultracold gases and Feshbach resonances are discussed.

- After that, Chapter 3 describes the experimental platform to explore a quantum degenerate Fermi-Fermi-Bose mixture and to investigate ultracold heteronuclear Fermi-Fermi molecules. The discussion begins with the experimental concept, followed by the vacuum system and the atomic sources. After that I give an overview of the laser system for trapping and manipulating the atoms. Then the imaging system and the setup for magnetic trapping are described, which is followed by the presentation of an ultra stable current control for the magnetic Feshbach fields. After that I show the setup for radio-frequency and microwave manipulation and finally I explain the experimental control in real-time.

1. Introduction

- In Chapter 4 I present the first magneto-optical trapping of two different fermionic species, ^{6}Li and ^{40}K, and a bosonic species ^{87}Rb. This also demonstrates simultaneous trapping of three species in a magneto-optical trap ("triple MOT") for the first time. Optimization of the atom numbers in the triple MOT configuration is shown and, in addition, loading and trap losses are characterized. These results pave the way towards a quantum degenerate mixture of two different fermionic species. Parts of this Chapter are published in (Taglieber *et al.*, 2006).

- The first quantum degenerate mixture of two fermionic species, ^{6}Li and ^{40}K, is presented in Chapter 5. I will give an overview on the experimental way to achieve such a quantum degenerate Fermi-Fermi mixture and show crucial experimental challenges. The two fermions are sympathetically cooled by a bosonic ^{87}Rb gas. In addition, a quantum degenerate two-species Fermi-Fermi mixture coexisting with a BEC is realized, representing simultaneous degeneracy of three species for the first time. Furthermore, an increased cooling efficiency for ^{6}Li by ^{87}Rb in the presence of ^{40}K is observed, demonstrating catalytic cooling. Parts of this Chapter are published in (Taglieber *et al.*, 2008).

- Finally, Chapter 6 presents the first creation of ultracold bosonic heteronuclear molecules of two fermionic species, ^{6}Li and ^{40}K, by a magnetic field sweep across an interspecies s-wave Feshbach resonance. First, a crossed optical dipole trap is characterized and the loading sequence of the atoms into this trap is described. After that, magnetic field calibration of the Feshbach coils and state preparation of ^{6}Li and ^{40}K to the desired states are explained. The discussion is followed by identifying Feshbach resonances and ends with measurements on heteronuclear Fermi-Fermi molecules. Parts of this chapter are published in (Voigt *et al.*, 2009).

Chapter 2

Theory

2.1 Ultracold gases

This theoretical part briefly summarizes the quantum statistical and thermodynamic properties of bosons and fermions. In the following, the main formulas for Bose-Einstein condensation and degenerate fermions are presented that allow to extract the essential physical parameters in an absorption measurement.

A detailed description can be found in several textbooks, e.g. (Cohen-Tannoudji et al., 1977a,b; Sakurai, 1994; Huang, 1987; Pitaevskii and Stringari, 2003).

2.1.1 Quantum statistical properties

In this chapter the basic quantum statistical and thermodynamic properties of bosons and fermions are discussed.

The total quantum mechanical wavefunction of identical particles is either symmetric or anti-symmetric under exchange of two particles. This leads to the classification of bosonic and fermionic particles for the symmetric or anti-symmetric case, respectively. Given the thermodynamic partition function for non-interacting particles, the mean occupation number $f(\epsilon_r)$ of a single particle energy eigenstate with energy ϵ_r can be deduced and is given by

$$f(\epsilon_r) = \frac{1}{e^{\beta(\epsilon_r - \mu)} + \theta} \text{, with } \theta \equiv \begin{cases} +1 & \text{Fermi-Dirac statistics} \\ -1 & \text{Bose-Einstein statistics} \\ 0 & \text{Maxwell-Boltzmann statistics} \end{cases}. \quad (2.1)$$

Here, $\beta = (k_B T)^{-1}$ specifies the temperature T of the system and μ is the chemical potential. The Fermi-Dirac statistics describes fermions, the Bose-Einstein statistics bosons and the Maxwell-Boltzmann statistics a classical system. For a given atom number N, the normalization condition of the sum over all possible states

$$N = \sum_r f(\epsilon_r) \quad (2.2)$$

fixes the chemical potential μ. Fig. 2.1 shows the chemical potential for bosons, fermions and distinguishable classical particles as a function of the temperature.

2. Theory

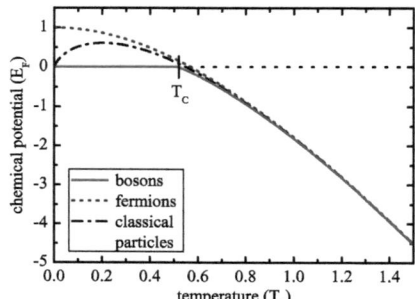

Figure 2.1: The chemical potential for bosons, fermions and distinguishable classical particles vs. temperature in a harmonic confinement. The critical temperature T_c for BEC is related to T_F by $T_c/T_F = (6\,g_3(1))^{-1/3} \approx 0.52$.

In experiments, the atomic ensemble is typically held in a harmonic potential $V_{\text{ho}}(\mathbf{r})$. This case is discussed and assumed in the following. A particle of mass m experiences the potential

$$V_{\text{ho}}(\mathbf{r}) = \frac{m}{2}\left(\omega_x^2 x^2 + \omega_y^2 y^2 + \omega_z^2 z^2\right), \qquad (2.3)$$

where ω_i ($i = x, y, z$) are the angular trapping frequencies. The mean trapping frequency is then given by $\bar{\omega} = (\omega_x \omega_y \omega_z)^{1/3}$. For large atom number and when the level spacing greatly exceeds the thermal energy $k_B T \gg \hbar \omega_i$ ($i = x, y, z$), the discrete sum in Eq. (2.2) can be replaced by an integral over the product between the mean occupation number $f(\epsilon)$ and the density of states $g(\epsilon)$. For harmonic trapping potentials the density of states is given by

$$g(\epsilon) = \frac{\epsilon^2}{2\,(\hbar\bar{\omega})^3}, \qquad (2.4)$$

which is exact in the thermodynamic limit. This expression fails for bosons at low temperatures, where the ground state is macroscopically occupied, but is weighted to zero. Therefore, the ground state for bosons has to be taken separately into account. On the other hand, identical fermions can only occupy at most one state and the ground state can be neglected for large atom numbers. The number of atoms in excited states is then given by

$$N_{\text{ex}} = \int_0^\infty f(\epsilon)\,g(\epsilon)\,d\epsilon. \qquad (2.5)$$

The integration can be calculated analytically and gives for the bosonic and fermionic cases

$$N_{\text{ex}} = -\theta \left(\frac{k_B T}{\hbar\bar{\omega}}\right)^3 g_3(-\theta\,\tilde{z}). \qquad (2.6)$$

Here, $\tilde{z} = e^{\beta\mu}$ defines the fugacity parameter. $g_\alpha(s)$ is the polylogarithm function and is given by the relation

$$g_\alpha(s) = \sum_{k=1}^\infty \frac{s^k}{k^\alpha}. \qquad (2.7)$$

It is defined for $\alpha, s \in \mathbb{C}$, $|s| < 1$ and its analytic continuation.

2.1 Ultracold gases

2.1.2 Bosons

2.1.2.1 Bose-Einstein condensate

The number of bosonic atoms for given temperature and trapping parameters in the excited states is a finite value. The system accommodates more particles in excited states as μ approaches zero. At this point additional atoms macroscopically occupy the ground state, while the chemical potential stays locked to zero. This phenomenon is called Bose-Einstein condensation (BEC). If N is the total atom number, then the atom number of the BEC is given by

$$N_0 = N - N_{\text{ex}}. \tag{2.8}$$

For non-interacting particles the critical temperature T_c for a BEC can be deduced from Eq. (2.6) by setting $N_0 = 0$ and $\mu = 0$:

$$T_c = \frac{\hbar \bar{\omega}}{k_B} \left(\frac{N}{g_3(1)} \right)^{1/3} \approx 0.94 \frac{\hbar \bar{\omega}}{k_B} N^{1/3}. \tag{2.9}$$

The condensate fraction is a function of temperature and for $T < T_c$ the fraction, which can be deduced from Eq. (2.6) and (2.9), is given by

$$\frac{N_0}{N} = 1 - \left(\frac{T}{T_c} \right)^3. \tag{2.10}$$

The density distribution of the trapped gas is calculated in Sec. 2.1.2.3 below. From Eq. (2.20) follows that BEC occurs if the peak density of the distribution fulfills the following condition:

$$n_0 \lambda_{\text{dB}}^3 = g_{3/2}(1) \approx 2.612, \tag{2.11}$$

where

$$\lambda_{\text{dB}} = \sqrt{\frac{2\pi \hbar^2}{m k_B T}} \tag{2.12}$$

is the thermal de Broglie wavelength. This means that BEC occurs when λ_{dB} is on the order of the mean particle separation and the indistinguishability of the particles is thus shown.

2.1.2.2 Weakly interacting Bose-Einstein condensate

Ultracold and dilute gases interact via elastic interaction and only binary interactions are relevant. At low temperatures collisions between particles occur in the s-wave scattering limit. The interaction can be written by a zero-range pseudo-potential $V(\mathbf{r} - \mathbf{r}') = g\delta(\mathbf{r} - \mathbf{r}')$ (Huang, 1987), where

$$g = \frac{4\pi \hbar^2 a}{m} \tag{2.13}$$

describes the coupling constant. In this equation a is the s-wave scattering length (see also discussion in Sec. 2.2.1). For [87]Rb the interaction is repulsive and the triplet scattering length is $106(4)\,a_0$.

In a mean field description the condensate wavefunction can be described by a macroscopic wavefunction $\Psi(\mathbf{r}, t)$, which has the meaning of an order parameter. The density of the condensate is then given by $n_c(\mathbf{r}, t) = |\Psi(\mathbf{r}, t)|^2$. In a variational approach, where a minimization

11

2. Theory

of the energy is performed, the Gross-Pitaevskii equation (GPE) can be derived. The time-independent GPE (Goldman and Silvera, 1981; Huse and Siggia, 1982) is given by

$$\mu \Psi(\mathbf{r}) = \left(-\frac{\hbar^2 \nabla^2}{2m} + V(\mathbf{r}) + g\left|\Psi(\mathbf{r})\right|^2\right) \Psi(\mathbf{r}) . \tag{2.14}$$

The equation is valid for temperatures $T \ll T_c$, large atom number $N_0 \gg 1$ and weakly interacting particles with $n_c |a|^3 \ll 1$. If $g n_c(\mathbf{r}) \gg \hbar \omega_i$ ($i = x, y, z$) for repulsive interactions (i.e $a > 0$), one can neglect the kinetic term of Eq. (2.14). This limit called the Thomas-Fermi approximation. The condensate density distribution is then given by

$$n_c(\mathbf{r}) = \max\left(\frac{\mu - V(\mathbf{r})}{g}, 0\right) . \tag{2.15}$$

For harmonic potentials the density distribution has a parabolic shape and vanishes when the chemical potential is equal to the trapping potential. The Thomas-Fermi radius of the condensate in the i direction is

$$R_i = \sqrt{\frac{2\mu}{m\omega_i^2}} . \tag{2.16}$$

The atom number of the condensate can be calculated by integrating over $n_c(\mathbf{r})$ and is given by

$$N_0 = \left(\frac{2\mu}{\hbar \bar{\omega}}\right)^{5/2} \frac{a_{\mathrm{ho}}}{15\, a} . \tag{2.17}$$

Here, $a_{\mathrm{ho}} = \sqrt{\hbar / m \bar{\omega}}$ is the oscillator length.

2.1.2.3 Density distribution and free expansion

Non-condensed gas The density distribution of the non-condensed part of the bosonic gas can be obtained by a semiclassical approximation of the phase space density distribution $\tilde{\rho}(\mathbf{r}, \mathbf{p})$. Here, a wave packet with definite position and momentum is assigned to each particle. The phase space density distribution is given by

$$\tilde{\rho}(\mathbf{r}, \mathbf{p}) = \frac{1}{h^3} \frac{1}{e^{\beta(p^2/2m + V(\mathbf{r}) - \mu)} - 1} . \tag{2.18}$$

In the experiment atomic clouds are typically probed after they are released from the trapping potential. For an arbitrary free expansion time the density of the thermal component can easily be deduced. For a harmonic trap it is given by

$$n(\mathbf{r}, t) = \int \int \tilde{\rho}(\mathbf{r}_0, \mathbf{p}) \delta^3\left(\mathbf{r} - \mathbf{r}_0 - \frac{\mathbf{p}\,t}{m}\right) d^3\mathbf{r}_0 d^3\mathbf{p} \tag{2.19}$$

$$= \frac{\prod_i \eta_i(t)}{\lambda_{\mathrm{dB}}^3} g_{3/2}\left(\tilde{z}\, e^{-\frac{m}{2k_\mathrm{B} T} \sum_i [\omega_i\, r_i\, \eta_i(t)]^2}\right) . \tag{2.20}$$

In this equation $\eta_i(t) = (1 + \omega_i^2 t^2)^{-1/2}$.

2.1 Ultracold gases

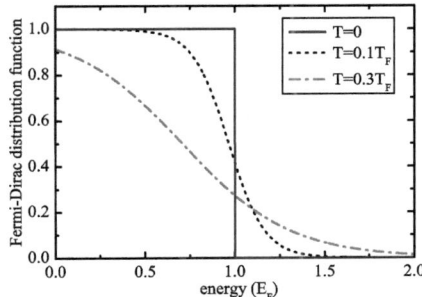

Figure 2.2: The Fermi-Dirac distribution function vs. energy for three different temperatures. In the $T \to 0$ limit, all energy states are occupied up to the Fermi energy $E_F = \mu(T \to 0, N)$.

Bose-Einstein condensate The expansion of a Bose-Einstein condensate in a radial symmetric, cigar-shaped trap evolves differently in comparison to the thermal part. The temporal evolution can be calculated from the GPE (Castin and Dum, 1996; Kagan et al., 1996; Dalfovo et al., 1997). The trapping frequencies are assumed to be ω_ρ along the radial direction and ω_z along the axial direction. Introducing the dimensionless parameters $\tau = \omega_\rho t$ and $\lambda = \omega_\rho/\omega_z$, the Thomas-Fermi radii in radial and axial direction evolve as

$$R_\rho(t) = R_\rho(0)\sqrt{1+\tau^2} \qquad (2.21)$$
$$R_z(t) = R_z(0)\left[1 + \lambda^{-2}\left(\tau \arctan \tau - \ln\sqrt{1+\tau^2}\right)\right]. \qquad (2.22)$$

These two equations show an anisotropic expansion from a cigar-shaped to a pancake-shaped cloud.

2.1.3 Fermions

2.1.3.1 Fermi energy

Spin-polarized fermions can only occupy one state, which is a consequence of the Pauli exclusion principle. When a Fermi gas is cooled towards lower temperatures, the Fermi-Dirac distribution function f_{FD} becomes relevant. For $T \to 0$ all energy states are filled up to the Fermi energy $E_F = \mu(T \to 0, N)$, which is defined by the chemical potential. The corresponding Fermi temperature is given by $T_F = E_F/k_B$. Fig. 2.2 shows the Fermi-Dirac distribution function for different temperatures. In the $T \to 0$ limit, f_{FD} evolves into a Heaviside step function that is 1 for temperatures below T_F and 0 above it.

The Fermi energy can be derived for a harmonic trap by Eq. (2.5) and gives

$$E_F = \hbar\bar{\omega}\left(6N\right)^{1/3}. \qquad (2.23)$$

In an experimental sequence the temperature of a Fermi gas is typically extracted in free

2. Theory

expansion. Here, the relation between the temperature and the fugacity is useful:

$$\frac{T}{T_{\rm F}} = \left(\frac{-1}{6\,g_3\,(-\tilde{z})}\right)^{1/3}. \tag{2.24}$$

This equation can be deduced from Eq. (2.6).

2.1.3.2 Density distribution and free expansion

The density distribution of a fermionic cloud can be derived from the phase space density distribution, as in the bosonic case. For details see also (Pitaevskii and Stringari, 2003). A Fermi gas at $T = 0$ has the following density distribution:

$$n(\mathbf{r},\, T=0) = \frac{(2\,m)^{3/2}}{6\,\pi^2\,\hbar^3}\,(E_{\rm F} - V(\mathbf{r}))^{3/2}$$

and is zero for $E_{\rm F} < V(\mathbf{r})$. For a harmonic trap the cloud size in the i direction can be described by the Fermi radius

$$R_i = \sqrt{\frac{2\,E_{\rm F}}{m\,\omega_i^2}}. \tag{2.25}$$

In a free expansion measurement the density distribution for an arbitrary time-of-flight and temperature is given by

$$n(\mathbf{r},\, t) = -\frac{\prod_i \eta_i(t)}{\lambda_{\rm dB}^3}\,g_{3/2}\left(-\tilde{z}\,e^{-\frac{m}{2k_{\rm B}T}\sum_i[\omega_i\,r_i\,\eta_i(t)]^2}\right). \tag{2.26}$$

This density distribution for fermions and the one for bosons, which is given by Eq. (2.20), differ only by the minus signs.

2.2 Feshbach resonances

The second theoretical part gives describes the physics of ultracold collisions in dilute gases and briefly discusses Feshbach resonances.

2.2.1 Ultracold Collisions

This section presents the basic principles of elastic collisions in ultracold, dilute gases. It shows that in the zero energy limit the scattering process can be described by one single parameter, the s-wave scattering length.

For details I refer to the following literature (Taylor, 1972; Landau and Lifschitz, 1985; Sakurai, 1994; Dalibard, 1999; Pitaevskii and Stringari, 2003).

2.2.1.1 Elastic collisions

In ultracold, dilute systems the thermal de Broglie wavelength $\lambda_{\rm dB}$ and the interparticle distance $n^{-1/3}$ are typically much larger than the range of the interatomic potential r_0, which is on the order of the van der Waals length $l_{\rm vdW}$. Therefore, for the description of the scattering process at a potential $V(\mathbf{r})$, the short-range scattering potential can be neglected.

The elastic scattering of an incoming plane wave with wavevector **k** can be described by the time-independent Schrödinger equation in the center-of-mass coordinates:

$$\left(-\frac{\hbar^2 \nabla^2}{2 m_r} + V(\mathbf{r})\right) \Psi_\mathbf{k}(\mathbf{r}) = E\, \Psi_\mathbf{k}(\mathbf{r}) \,. \tag{2.27}$$

Here, $m_r = m_1 m_2 / (m_1 + m_2)$ is the reduced mass and $E = \hbar^2 k^2 / (2 m_r)$ is the energy of the plane wave. For large distances, $r \to \infty$, the wavefunction $\Psi_\mathbf{k}(\mathbf{r})$ possesses the asymptotic form

$$\Psi_\mathbf{k}(\mathbf{r}) \sim e^{i\mathbf{k}\cdot\mathbf{r}} + f(k, \vartheta, \varphi)\, \frac{e^{ikr}}{r}, \tag{2.28}$$

with the spherical coordinates r, ϑ, φ. This expression corresponds to a sum of an incoming plane wave and an outgoing scattered part that is described by a scattering amplitude $f(k, \vartheta, \varphi)$. The cross section is given by

$$\sigma(k) = \int_\Omega |f(k, \vartheta, \varphi)|^2\, d\Omega\,. \tag{2.29}$$

For a central potential one can perform the standard expansion of the incident and outgoing wave into partial waves with angular momenta l. Here, l stands for s-, p-, d-, ... waves. Scattering at a central potential does not change l, but induces a phase shift δ_l. This allows to express the cross section $\sigma(k)$ as a sum over all partial wave cross sections (Landau and Lifschitz, 1985)

$$\sigma(k) = \sum_{l=0}^\infty \sigma_l(k)\,, \text{ with } \sigma_l(k) = \frac{4\pi}{k^2}(2l+1)\sin^2(\delta_l)\,. \tag{2.30}$$

2.2.1.2 Low-energy scattering (Zero-energy limit)

In ultracold gases with temperatures of up to several 100 μK the scattering process is described by low momenta $k \ll 1/r_0$. The scattering phase *modulo* 2π scales as $\delta_l \propto k^{2l+1}$ for $k \to 0$ (Landau and Lifschitz, 1985). Thus, the partial wave cross section with angular momenta l has the scaling

$$\sigma_l \propto k^{4l} \propto E^{2l} \tag{2.31}$$

for $k \to 0$. This shows, that in the low temperature limit, the s-wave collisions are dominant and collisions of angular momentum $l \geq 1$ are strongly suppressed. For $k \ll 1/r_0$ the effective-range expansion to second order in k gives (Landau and Lifschitz, 1985)

$$k \cot \delta_0 \approx -\frac{1}{a} + r_\text{eff}\frac{k^2}{2} \tag{2.32}$$

with

$$a = -\lim_{k \to 0} \frac{\tan \delta_0(k)}{k}, \tag{2.33}$$

which defines the effective range r_eff of the scattering potential and the scattering length a. The effective range r_eff for atom-atom interactions with large interatomic separations, that can be

2. Theory

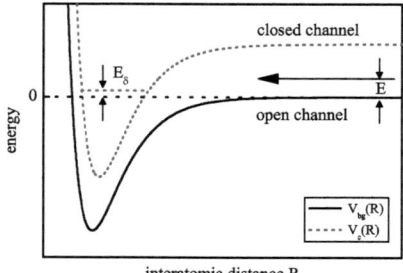

Figure 2.3: The graph shows two molecular potentials $V_{bg}(R)$ and $V_c(R)$ vs. the interatomic distance. E_δ is the threshold energy difference between a molecular bound state and the open channel. If the open and closed channel couple to each other, the bound molecular state can be occupied and the scattering is resonantly enhanced.

described by the van der Waals potential $V(r) = -C_6/r^6$, is on the order of the van der Waals length l_{vdW} (Flambaum *et al.*, 1999) that is given by

$$l_{vdW} = \frac{1}{2}\left(\frac{2\,m_r\,C_6}{\hbar^2}\right)^{1/4}. \tag{2.34}$$

In general, the scattering length a is not restricted and can be $-\infty < a < +\infty$. With the above expansion the scattering amplitude can be expressed as

$$f(k) = \frac{1}{-\frac{1}{a} + \frac{1}{2}r_{\text{eff}}\,k^2 - i\,k}. \tag{2.35}$$

In the case of $k\,|a| \gg 1$ and $r_{\text{eff}} \ll 1/k$ the scattering amplitude depends only on momentum and yields $f = i/k$. This regime is called the unitarity limit and the cross section of distinguishable particles is $\sigma = 4\pi/k^2$.

For $k\,|a| \ll 1$ and $|r_{\text{eff}}| \lesssim 1/k$ the scattering amplitude is a function of one single parameter, $f = -a$, and is independent on momentum. In this zero energy limit the scattering cross section of distinguishable particles is $\sigma = 4\pi a^2$. For identical particles the scattering amplitudes interfere and the total wavefunction has to be symmetrized or anti-symmetrized. This leads to doubling of the symmetric even partial waves contributions for the bosons and cancelling of the anti-symmetric odd partial waves. For fermions, on the other hand, the odd partial waves contributions are doubled and the even ones are cancelled. Therefore, identical fermions do not scatter in the zero energy limit and form an ideal gas. Summarized, the s-wave cross section is given by

$$\sigma = \begin{cases} 4\pi a^2 & \text{distinguishable particles} \\ 8\pi a^2 & \text{identical bosons} \\ 0 & \text{identical fermions} \end{cases}. \tag{2.36}$$

2.2 Feshbach resonances

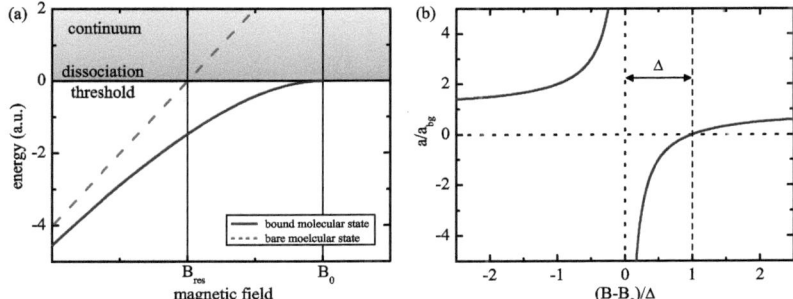

Figure 2.4: (a) In the vicinity of a Feshbach resonance the coupled, bound molecular state and the uncoupled, bare molecular state is plotted vs. the magnetic field. Due to interchannel coupling, the resonance position of the dressed molecular state B_0 is shifted with respect to the threshold crossing B_{res} of the bare, uncoupled state. The shaded area above the dissociation threshold represents the continuum. (b) The scattering length is plotted as a function of the magnetic field. At the resonance position the scattering length has a pole, $a \to \pm\infty$.

2.2.2 Magnetic Feshbach resonance

The collision process, so far discussed, contains only a single interaction potential. In general, binary collisions in ultracold gases are more complex due to different internal states of the particles. One possible consequence is that the collision partners might get resonant to a molecular bound state and form a Feshbach resonance (Fano, 1935, 1961; Feshbach, 1958, 1962). In ultracold quantum gases the Feshbach resonance is a key tool for many breakthrough experiments. With a Feshbach resonance the interactions of quantum matter can be controlled precisely and the study many-body physics and condensed matter phenomena are possible. Moreover, molecular physics in the zero temperature limit is accessible.

In the following the basic principle of a magnetically induced Feshbach resonance is discussed. For details I refer to the review articles (Köhler *et al.*, 2006; Chin *et al.*, in preparation).

2.2.2.1 Principle

A simple model of the Feshbach resonance takes two molecular potentials $V_{\text{bg}}(R)$ and $V_{\text{c}}(R)$ into account. Both of them have a different threshold energy. For small kinetic energies E two free atoms with large interatomic distance approach each other in the so-called entrance channel or open channel and enter the background potential $V_{\text{bg}}(R)$. Due to the small kinetic energy the upper potential $V_{\text{c}}(R)$ is energetically closed and is called the closed channel.

This closed channel can have bound molecular states that are near the threshold or dissociation energy of the open channel. If the open and closed channel couple to each other, the bound molecular state can be occupied and the scattering is resonantly enhanced. The molecular potentials are illustrated in Fig. 2.3.

In general the magnetic moments of the bare bound molecular state μ_{mol} and of the asymptotically separated atoms μ_{atoms} are different. μ_{res} is defined as the difference between these

2. Theory

two magnetic moments. Then the energy difference of the two states, given to first order by

$$E_\delta = \mu_{\text{res}} \left(B - B_{\text{res}}\right), \tag{2.37}$$

can easily be varied by an external magnetic field of strength B. In the above expression the magnetic field B_{res} is the threshold crossing of the bare, uncoupled state. This tunability leads to magnetically induced Feshbach resonances, which allows a precise control of the scattering length. For s-wave collisions the scattering length has a simple form and is given by (Moerdijk et al., 1995)

$$a(B) = a_{\text{bg}} \left(1 - \frac{\Delta}{B - B_0}\right). \tag{2.38}$$

This expression only depends on three parameters: a_{bg} is the background scattering length of the potential $V_{\text{bg}}(R)$, Δ is the resonance width and B_0 is the resonance position of the dressed molecular state, where $a(B)$ diverges. B_0 is shifted with respect to B_{res} due to interchannel coupling and the shift is given by the relation (Köhler et al., 2006)

$$B_0 - B_{\text{res}} = \Delta \frac{a_{\text{bg}}}{\bar{a}} \frac{1 - a_{\text{bg}}/\bar{a}}{1 + (1 - a_{\text{bg}}/\bar{a})^2}. \tag{2.39}$$

Here, $\bar{a}$ defines a mean scattering length that is defined as

$$\bar{a} = \frac{4\pi}{\Gamma(1/4)^2} l_{\text{vdW}} \approx 0.95598 \, l_{\text{vdW}}. \tag{2.40}$$

Fig. 2.4 shows the molecular states and the scattering length dependency on the magnetic field.

2.2.3 The asymptotic bound state model

A realistic model for the interaction of two alkali atoms includes the relative electronic spin orientation. In this section, the asymptotic bound state model (Wille et al., 2008) is discussed, which is based on the models by (Moerdijk et al., 1995) and (Stan et al., 2004) and is extended by a mixing term between singlet and triplet states (see also (Walraven, 2009)). The coupled-channel calculation (Stoof et al., 1988), which is computationally more demanding, takes the exact interaction potentials and the coupling between spin channels into account.

In the Born-Oppenheimer approximation the electronic motion is effectively decoupled from the nuclear motion. At ultracold temperatures the molecule is in its electronic ground state potential and has a zero angular momentum. The electronic spins of the two colliding alkali atoms are in a S-state and are either in the singlet or triplet state. The relevant singlet potential $V_{s=0}(r)$ is then $X^1\Sigma^+$ and the triplet potential $V_{s=1}(r)$ is $a^3\Sigma^+$. These potentials are isotropic, which allows to write the relative motion of the atoms through the following Hamiltonian:

$$H^{\text{rel}} = \frac{\hbar^2}{2m_{\text{r}}} \left(-\frac{d^2}{dr^2} + \frac{l(l+1)}{r^2}\right) + \sum_{S=0,1} V_s(r) P_s. \tag{2.41}$$

Here l is the projection quantum number of the angular momentum, m_{r} is the reduced mass, r is the interatomic separation and P_s is the projection operator onto the singlet or triplet spin states.

2.2 Feshbach resonances

The total Hamiltonian that includes the hyperfine structure is given by

$$H = H^{\text{hf}} + H^{Z} + H^{\text{rel}}. \tag{2.42}$$

The two first terms represent the hyperfine and Zeeman energies of each atom and are given by

$$H^{\text{hf}} = \frac{a_1^{\text{hf}}}{\hbar^2} \mathbf{s}_1 \cdot \mathbf{i}_1 + \frac{a_2^{\text{hf}}}{\hbar^2} \mathbf{s}_2 \cdot \mathbf{i}_2 \tag{2.43}$$

$$H^{Z} = \gamma^e \mathbf{S} \cdot \mathbf{B} - \gamma_1^n \mathbf{i}_1 \cdot \mathbf{B} - \gamma_2^n \mathbf{i}_2 \cdot \mathbf{B}, \tag{2.44}$$

where $a_{\text{Li}}^{\text{hf}}$ and a_{K}^{hf} are the hyperfine coupling constants of ^{6}Li and ^{40}K. $\mathbf{s}$ is the single-atom electron spin and $\mathbf{i}$ is the corresponding nuclear spin. The total electronic spin of the two atoms is $\mathbf{S} = \mathbf{s}_1 + \mathbf{s}_2$. γ^e and γ^n are the gyromagnetic ratios of the electron and nuclei.

The hyperfine Hamiltonian H^{hf} can be rewritten in terms of $(\mathbf{s}_{\text{Li}} + \mathbf{s}_{\text{K}})$ and $(\mathbf{s}_{\text{Li}} - \mathbf{s}_{\text{K}})$ and gives

$$H^{\text{hf}} = H_+^{\text{hf}} + H_-^{\text{hf}} \tag{2.45}$$

with

$$H_+^{\text{hf}} = \frac{a_1^{\text{hf}}}{2\hbar^2} \mathbf{S} \cdot \mathbf{i}_1 + \frac{a_2^{\text{hf}}}{2\hbar^2} \mathbf{S} \cdot \mathbf{i}_2 \tag{2.46}$$

$$H_-^{\text{hf}} = \frac{a_1^{\text{hf}}}{2\hbar^2} (\mathbf{s}_1 - \mathbf{s}_2) \cdot \mathbf{i}_1 - \frac{a_2^{\text{hf}}}{2\hbar^2} (\mathbf{s}_1 - \mathbf{s}_2) \cdot \mathbf{i}_2. \tag{2.47}$$

The Hamiltonian H_+^{hf} contains terms proportional to the total electronic spin $\mathbf{S}$, and thus preserves the separation of the orbital from the spin problem. H_-^{hf}, however, mixes singlet and triplet states.

The full Hamiltonian of Eq. (2.42) conserves the quantum numbers l and the total angular momentum $M_{\text{F}} = M_S + m_1 + m_2$. The quantum numbers M_S, m_1 and m_2 belongs to the spins $\mathbf{S}$, $\mathbf{s}_1$ and $\mathbf{s}_2$. Now, the asymptotic bound state model (Wille et al., 2008) expands the wavefunction $|\Psi^l\rangle$ of the Hamiltonian of Eq. (2.42) in terms of $|\psi_S^l\rangle |S, M_S, m_1, m_2\rangle$. The first part of this expansion $|\psi_S^l\rangle$ describes the asymptotic last bound state of $l(l+1)/(2 m_r r^2) + V_S$. The second part $|S, M_S, m_1, m_2\rangle$ describes the spin function.

In the ^{6}Li-^{40}K mixture the Franck-Condon factor of the asymptotically last bound state of the singlet and triplet potential is close to unity (Wille et al., 2008), i.e. $\langle \psi_{S=0}^l | \psi_{S=1}^l \rangle \simeq 1$. For this case the characteristic equation for the coupled bound state energies E can be approximated by

$$|\langle S', M_S', m_1', m_2'| H_+^{\text{hf}} + H_-^{\text{hf}} + H^Z + E_S^l - E|S, M_S, m_1, m_2\rangle| = 0. \tag{2.48}$$

Here E_S^l is the energy of the last bound state with angular momentum l and spin S. The energy $E_S^{l=0}$ corresponds to the singlet and triplet scattering lengths. It can be shown, that including additional information about the potential shape, the bound state energies E_S^l with $l > 0$ can be deduced from $E_S^{l=0}$ (Wille et al., 2008).

2.2.3.1 Feshbach resonances in the ^{6}Li-^{40}K mixture

The ^{6}Li-^{40}K mixture offers several s- and p-wave Feshbach resonances. The last bound state energies $E_S^{l=0}$ can be derived by fitting the observed resonances with the solutions of Eq. (2.48).

2. Theory

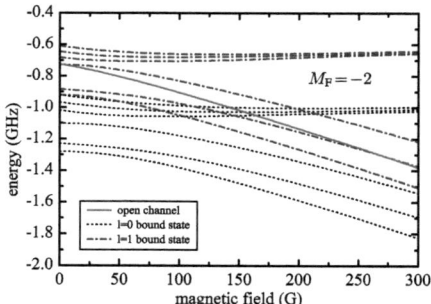

Figure 2.5: The bound state energies for the $M_F = -2$ collision channel between ^{6}Li $|F = 1/2, m_F = 1/2\rangle$ and ^{40}K $|F = 9/2, m_F = -5/2\rangle$. The bound state energies for s- and p-waves ($l = 0$ and $l = 1$) and the open channel are plotted versus the magnetic field. The crossings between the molecular bound states and the open channel give the position of the Feshbach resonances.

According to (Wille et al., 2008), the parameters are $E_{S=0}^{l=0} = h \times 716\,(15)$ MHz and $E_{S=1}^{l=0} = h \times 425\,(5)$ MHz.

As an example the bound state energies for the $M_F = -2$ collision channel between ^{6}Li $|F = 1/2, m_F = 1/2\rangle$ and ^{40}K $|F = 9/2, m_F = -5/2\rangle$ are calculated. Fig. 2.5 shows the molecular bound state energies for s- and p-waves and the open channel.

The crossings between the molecular bound states and the open channel give the position of the Feshbach resonances. In the $M_F = -2$ case three s-wave and one p-wave Feshbach resonances are experimentally observed. The asymptotic bound state model predicts one additional, not observed p-wave resonance at 17.5 G. The group of Prof. E. Tiemann calculated the resonances with a more accurate model recently and reported, that this resonance does not appear (Tiemann et al., 2009).

2.2.4 Classification of Feshbach resonances

The physics at a Feshbach resonance is in principle affected by the open and closed channel states. However, it can be shown that for certain parameters the closed channel contributions can be neglected and, thus, the physical description of the system is independent of the properties of the molecular state. In this sense, the system is suitable to realize universal many-body phenomena. In the following, the conditions of universal behavior are discussed for fermionic systems (for details see (Sheehy and Radzihovsky, 2007; Ketterle and Zwierlein, 2008)).

The relevant energy scales are the Fermi energy E_F, the energy difference E_δ of the open and closed channel and finally the coupling energy E_0 of the Feshbach resonance. E_0 can be expressed by experimental measurable observables (Ketterle and Zwierlein, 2008):

$$E_0 = \frac{1}{2} \frac{(\mu_{\text{res}} \Delta)^2}{\hbar^2/m\, a_{\text{bg}}^2}. \tag{2.49}$$

On the molecular or BEC side of the Feshbach resonance ($a > 0$) the dressed molecular state is a superposition of the closed and open channel components. The fraction $Z(B)$ of the closed channel part can be related to the difference in magnetic moments of the two channels (Duine and Stoof, 2003; Köhler et al., 2006):

$$Z(B) = \frac{1}{\mu_{\text{res}}} \frac{\partial E_{\text{b}}}{\partial B}. \qquad (2.50)$$

Here E_{b} is the binding energy of the bound molecular state. Close to resonance, in the universal regime, the binding energy of the molecule depends solely on the scattering length a (Köhler et al., 2006):

$$E_{\text{b}} = -\frac{\hbar^2}{m\,a^2}. \qquad (2.51)$$

In this regime the molecule forms a "halo" dimer with a bond length of $\langle r \rangle = a/2$ that greatly exceeds the outer classical turning point for $a \gg l_{\text{vdW}}$. Combining Eq. (2.50) and (2.51) gives together with Eq. (2.49) and the Fermi energy E_{F} the following condition:

$$Z(B) = 2\sqrt{\frac{E_{\text{F}}}{E_0}}\frac{1}{k_{\text{F}}\,a}. \qquad (2.52)$$

The condition for a negligible closed channel admixture in terms of the magnetic field range $|B - B_0|$ is given by the relation (Köhler et al., 2006)

$$\left|\frac{B - B_0}{\Delta}\right| \ll \frac{|\mu_{\text{res}}\Delta|}{2\,\hbar^2/m\,a_{\text{bg}}^2}. \qquad (2.53)$$

On the atomic or BCS side of the Feshbach resonance ($a < 0$) the system is still affected by the closed channel even though the bound molecular state disappears. A finite-energy resonance appears for $E_\delta > E_0$ in the scattering cross-section (Andreev et al., 2004; Gurarie and Radzihovsky, 2007). This resonant state has a peak energy at

$$E_* = E_\delta - E_0 \qquad (2.54)$$

and finite lifetime $\hbar\,\Gamma_*^{-1}$ with

$$\Gamma_* = 2\,E_0\sqrt{\frac{2\,E_\delta}{E_0} - 1}. \qquad (2.55)$$

Thus, scattering at the finite-energy resonance is non-universal.

On resonance the effective range of the scattering potential has the value (Ketterle and Zwierlein, 2008)

$$r_{\text{eff}} = -\sqrt{\frac{2\,\hbar^2}{m\,E_0}} \qquad (2.56)$$

and effects the scattering amplitude of Eq. (2.35) by the $\frac{1}{2}\,r_{\text{eff}}\,k^2$ contribution.

For universal behavior the physics should not depend on the molecular state for energies $|E_\delta| \leq E_{\text{F}}$. Furthermore, molecular effects throughout the entire strongly interacting regime $k_{\text{F}}\,|a| \geq 1$ should be negligible. Applying these two constraints on the above three regimes, one gets the universality condition $E_0 \gg E_{\text{F}}$ or equivalently $k_{\text{F}}\,r_{\text{eff}} \ll 1$. This means that the interparticle separation has to be much larger than the effective range of the scattering potential. Such a Feshbach resonance is called broad resonance, because E_0 scales quadratically with the resonance width Δ and the condition is usually fulfilled for resonances much larger than 1 G.

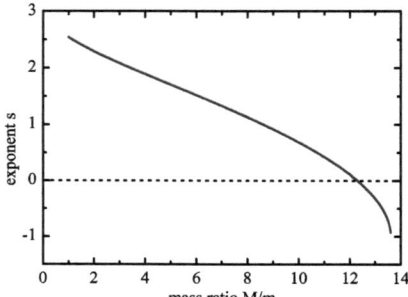

Figure 2.6: The molecular relaxation rate α_{rel} is strongly affected by quantum statistics. In fermionic systems the relaxation rate scales as $\alpha_{\text{rel}} \propto a^{-s}$. The figure shows the exponent s for dimer-dimer collisions as a function of the mass ratio M/m of the fermionic atoms.

In the other case, $E_0 \ll E_F$, the resonance is called narrow and the molecular state can not be neglected and has to be included in the many-body description (De Palo et al., 2004; Simonucci et al., 2005; Sheehy and Radzihovsky, 2007). At such a narrow resonance molecules in the closed channel state can already form on the BCS side ($a < 0$) for $E_\delta > 0$ (Falco and Stoof, 2004; Gurarie and Radzihovsky, 2007). Due to equilibrium of the fermions and molecules the chemical potential holds the condition $2\mu \leq E_\delta$. Hence, it is energetically favorable to convert a certain fraction of the fermions into molecules in the regime $0 < E_\delta < 2E_F$. It is expected that the decay of these molecules into free atoms is Pauli blocked due to the presence of the Fermi sea (Falco and Stoof, 2004).

2.2.5 Fermi-Fermi mixtures

The collisional properties of Feshbach molecules crucially depend on the quantum statistical properties of the corresponding constituents. For Fermi-Fermi molecules a surprisingly long lifetime at high densities of about 10^{13} cm^{-3} was observed in the vicinity of a Feshbach resonance (Cubizolles et al., 2003; Jochim et al., 2003a; Strecker et al., 2003; Regal et al., 2004a). The molecular lifetime was about 100 ms for ^{40}K and several seconds for ^{6}Li. The mechanism, explained by (Petrov et al., 2004) for halo dimers, is briefly discussed in the following.

The increased lifetime is attributed to a suppression of inelastic decay by vibrational quenching to lower vibrational molecular states. The corresponding mechanism is related to two facts: First, the wavefunction overlap between a Feshbach molecule with a large spatial extension and more deeply bound molecular states is small; second, Pauli suppression holds for the collisional process between fermions. A halo dimer has a characteristic size that is given by the scattering length a. The spatial size of deeply bound molecular states is r_0, which is on the order of the van der Waals length l_{vdW} and is much smaller than a close to resonance. A relaxation to deeply bound states requires at least three atoms that come close at a distance r_0. Since two of the three fermionic atoms are necessarily identical, the relative wavefunction has to be antisymmetric. Consequently, the wavefunction has a node at the relative distance $r = 0$ and varies for small values of r as $\sim kr$. The characteristic momentum spread of the atoms of the halo dimer is $k \sim 1/a$. Therefore, the relaxation probability is suppressed by a certain power of $(k r_0) \sim (r_0/a)$.

The exact relaxation rate α_{rel} has been first calculated for equal masses (Petrov et al., 2004) and has been extended for unequal ones (Petrov et al., 2005). The derivation assumes a negligible component of the closed channel part and a binding energy that obeys the universal scaling with the scattering length a according to Eq. (2.51). α_{rel} is found to be

$$\alpha_{\text{rel}} = C \frac{\hbar r_0}{m} \left(\frac{r_0}{a}\right)^s . \tag{2.57}$$

Here, the coefficient C is a system dependent parameter and s is an exponent that depends on the collisional partners. This exponent for dimer-dimer collisions is calculated according to (Petrov et al., 2005) and is plotted in Fig. 2.6. For equal masses $M = m$ of the fermionic atoms the exponent is $s = 2.55$ and continuously decreases for larger mass ratios. For ^{40}K and ^{6}Li $s = 1.39$ and is zero at a mass ratio of $M/m = 12.33$. For larger mass ratios the collisional relaxation increases up to $M/m = 13.6$. For larger mass ratios $M/m > 13.6$ a short-range three-body parameter is necessary for the calculation. However, for very large mass ratios $M/m > 100$ a gas-crystal quantum transition is predicted (Petrov et al., 2007).

In contrast to fermions, bosons show an increased relaxation rate at a Feshbach resonance that scales as a^4 with the scattering length (Fedichev et al., 1996b; Weber et al., 2003; Petrov, 2004), even though the wavefunction overlap between the initial halo dimer state and final deeply bound molecular states decreases.

2. Theory

Chapter 3

Experimental Setup

This chapter describes the experimental platform to explore a quantum degenerate Fermi-Fermi-Bose mixture and to investigate ultracold heteronuclear Fermi-Fermi molecules. The design and realization of this platform is one of the main parts of this thesis.

In the following sections I will describe the experimental apparatus in detail. The discussion begins with the experimental concept in Sec. 3.1, followed by the vacuum system in Sec. 3.2 and the atomic sources in Sec. 3.3. After that I give an overview of the laser system for trapping and manipulating the atoms, Sec. 3.4. The imaging system is described in Sec. 3.5, followed by the setup for magnetic trapping, Sec. 3.6. An ultra stable current control for the magnetic Feshbach fields is explained in Sec. 3.7. Then I show the setup for radio-frequency and microwave manipulation in Sec. 3.9 and finally I explain the experimental control in real-time in Sec. 3.10.

3.1 Experimental concept

The main idea of the experimental platform is to explore quantum degenerate mixtures of two atomic fermionic species. Such a Fermi-Fermi system is of particular interest in the vicinity of a Feshbach resonance. Here it is possible to create heteronuclear bosonic molecules, which can either be used to create dipolar molecules or to explore the BEC-BCS crossover.

In our experiment we decided to use the fermionic alkali atoms ^{6}Li and ^{40}K. For cooling into the quantum degenerate regime we chose the strategy by sympathetic cooling the two fermions by a cooling agent, namely the bosonic alkali atom ^{87}Rb. This avoids the reduction in atom number of the fermions by evaporation, and therefore the need for high number atom sources. Moreover, evaporation of fermions would have the disadvantage of Pauli blocking at the final stage of evaporation.

From the experimental point of view, the selected species have special advantages. All three species are well-known and have a simple energy structure. They have been cooled to quantum degeneracy in single species configuration already, but not together. After designing, building-up the system and working on quantum degeneracy, even the combination of ^{6}Li and ^{87}Rb have not been realized before. All species are alkali atoms and therefore one expects several interspecies Feshbach resonances. The necessary laser systems for trapping, cooling and manipulation can be built up by cost-effective semiconductor laser sources. The laser wavelengths are close together and allow to use common optics.

The concept of the experimental setup to achieve quantum degeneracy is as follows: First,

3. Experimental Setup

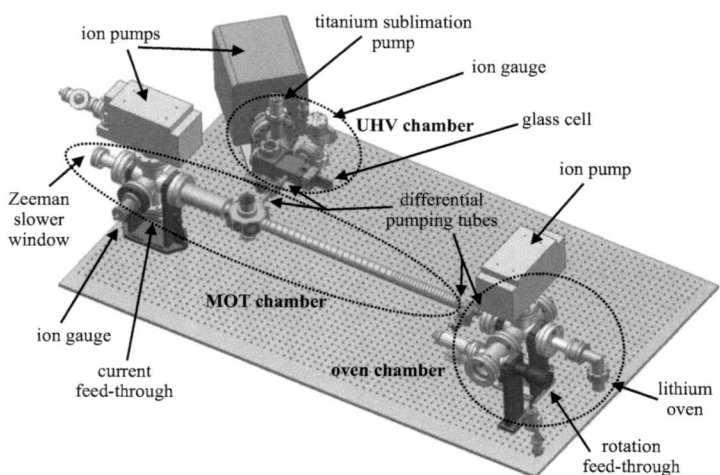

Figure 3.1: The vacuum system for experimental studies with the atomic species lithium, potassium and rubidium consists of three different parts: the oven chamber, the MOT chamber and the UHV chamber.

the atoms are loaded simultaneously into a magneto-optical trap (MOT) in a vacuum chamber ("MOT chamber"), lithium from a Zeeman slower, potassium and rubidium from background vapor pressure. To achieve longer lifetimes, necessary for evaporative and sympathetic cooling, the atoms are transferred into an ultra high vacuum chamber ("UHV chamber"). For excellent optical access, the transfer is realized by a magnetic transport (Greiner et al., 2001) around a corner into a glass cell. Here the atoms are transferred into a Ioffe-type trap to suppress Majorana losses. Through selective evaporative cooling of rubidium, the fermions are sympathetically cooled to quantum degeneracy. For studies with Feshbach resonances, the atoms are loaded into an optical dipole trap.

3.2 Vacuum system

For experiments with atomic quantum gases in magnetic traps based on coils ultra-high vacuum, typically below 1×10^{-11} mbar, is needed. This reduces losses and heating from background gas during the experimental cycle. However, for MOT loading from vapor dispensers high partial pressure is required. Therefore, the vacuum system consists of three different parts, the oven chamber for lithium, the MOT chamber and the UHV chamber (see Fig. 3.1).

The oven chamber is used for loading lithium with a Zeeman slower (Phillips and Metcalf, 1982) and gives a collimated atomic beam (for details to the Zeeman slower see the following Sec. 3.3). This chamber starts with a lithium oven that can be heated up to several 100 °C in a controlled way (for details see Voigt (2004)). The oven is connected to a five-way connector through a differential pumping stage (6 mm inner diameter, 23 cm length) to reduce the atomic flux into this chamber. The atomic beam can be blocked by an atomic beam shutter, connected to a rotation feed-through. For optical analysis of the atomic beam two windows are flanched to the five-way connector. The oven chamber is pumped by an 50 l/s ion pump (Varian, VacIon Plus 55 StarCell) and connects the MOT chamber through a second differential pumping stage, two tubes with an inner diameter of 6 mm and a length of 16.5 cm. Between the tubes, a pneumatically actuated valve interrupts the vacuum connection and allows refilling the lithium oven without breaking the whole vacuum system.

A steel tube for the Zeeman slower (77 cm long) connects the valve with the MOT chamber that has a flat octagonal shape with a height of 46 mm. This allows close placement of magnetic coils to the atomic clouds for efficient operation of the magnetic transport. The octagonal chamber has indium-sealed quartz windows with a broadband antireflection (AR) coating, one small window for optical pumping and six larger ones with a clear diameter of 40 mm that allow large MOT beams. In extension of the oven chamber, a second five-way connector is attached to the MOT chamber. To one end a window for the Zeeman slower beam is connected and is heated up to 165 °C to prevent coating of lithium atoms. Furthermore a 50 l/s ion pump (same type as above) guarantees low pressure in the MOT chamber and an ion gauge (Varian, UHV-24p) monitors the vacuum pressure. Finally, a six-channel electrical feedthrough (VTS Schwarz), which is connected from the bottom, is used for six atomic dispensers.

The MOT chamber is then connected to UHV chamber via two differential pumping tubes of 10 cm and 7.4 cm length and with an inner diameter of 8 mm. The tubes are intersected by a pneumatically actuated valve that allows opening the vacuum of the MOT chamber without affecting the vacuum in the UHV chamber or vice versa. Low pressure is achieved by a large ion pump, 125 l/s (Varian, VacIon Plus 150 StarCell), and a titanium sublimation pump (Thermionics) that increases the pumping speed for reactive, getterable gases like hydrogen and nitrogen. An additional ion gauge (same type as above) measures the vacuum pressure and is switched off during the experiment. In extension from the MOT chamber to the tubes a second window for optical pumping and detection is connected to the end of the UHV chamber. Perpendicular to this extension line a glass cell is attached to the chamber that allows, in combination with an additional window on the other side of the glass cell, excellent optical access along six axes. The glass cell is a quartz glass by Helma with a broadband AR coating ($R < 0.5\%$ for $512 - 1064$ nm at normal incidence). It consists of two parts, a rectangular part with outer dimensions of $26 \times 26 \times 70.5$ mm^3 and a wall thickness of 4 mm and a circular part of two stacked circular discs. The discs have an outer diameter of 37 mm and 50 mm, a length of 19 mm and a central hole of 18 mm. The glass cell is mounted to the steel chamber with a spring loaded sealing ring (Garlock, HNV 200 Helicoflex Delta) in-between.

After installing the vacuum system, all chambers are pumped down by an attached turbomolecular and membrane pump. To allow outgassing from the bulk material, the system is baked out at roughly 200 °C for several days. During bake-out, the atomic vapor dispensers and the filaments of the titanium sublimation pump are initialized. After gradually reducing the temperature, the ion pumps are switched on and a pressure of a few times 10^{-10} mbar is reached in the MOT chamber. The pressure in the UHV chamber is below the detection limit of 1×10^{-11} mbar.

3.3 Atomic sources

Due to different physical and chemical properties of ^{6}Li, ^{40}K and ^{87}Rb different strategies are used for loading the MOT. The species potassium and rubidium are loaded from background gas, provided by atomic vapor dispensers. For lithium this experimental simple loading scheme would be inefficient due to a much lower saturation pressure (see also Voigt (2004)) and a smaller mass resulting in a small fraction of atoms travelling at speeds below the MOT capture velocity. Therefore, lithium is loaded by a Zeeman slower, which requires extra magnetic coils and additional lasers.

3.3.1 Dispensers for Rb and K

^{40}K and ^{87}Rb are loaded from background gas into the MOT. The atoms are emitted from electrically heated dispensers through a redox reaction between a salt and a reducing agent. In total, three dispensers for each species are installed in the vacuum. The dispensers are mounted on two macor rings and placed closed to the MOT in hollows of the octagonal MOT chamber. Direct emission into the MOT is blocked by an additional shielding wire. The electrical connection of the dispensers allows to run one or two dispensers for each species. In the experiment, usually only two dispensers of each species are used to have one as a backup.

Due to a natural abundance of 28% for ^{87}Rb, commercial dispensers (SAES Getters, Rb/NF/7/25FT10+10) are used for loading ^{87}Rb into the MOT. For ^{40}K the natural abundance is only 0.0117(1) % and does not allow the application of commercial dispensers.

Therefore, dispensers have to be build using enriched potassium (DeMarco et al., 1999b), which relies on the following redox reaction:

$$2KCl + Ca \rightarrow 2K + CaCl_2 \quad (3.1)$$

In the first dispenser generation a ^{40}K abundance of 3% is used. A detailed description of the extensive manufacturing process can be found in (Henkel, 2005).

For an advanced second generation of ^{40}K dispensers the production process is improved and a larger abundance of 6% is used (MaTeck, Jülich). In this version, 10 mg of KCl and 20 mg of pure calcium (Sigma-Aldrich, pureness > 99.99%) are filled in three containers made of a nickel-chromium foil (Goodfellow Inc., Ni80/Cr20). For this 2nd generation, the whole production process is done under a protective atmosphere with dry argon. Furthermore, to significantly suppress impurities, all components are baked-out in a test vacuum setup before installing them in the MOT chamber. In addition, an improved shielding of the MOT region with thicker wires is installed.

3.3.2 Zeeman slower

Lithium is loaded into the MOT with the Zeeman slower technique, which was first demonstrated by (Phillips and Metcalf, 1982). The principle of this elegant and efficient technique relies on a combination of laser cooling (Hänsch and Schawlow, 1975) and the Zeeman effect. An atomic beam is slowed down and cooled by the light pressure of a resonant counter-propagating laser beam. To keep the atoms always in resonance with the laser light, a position-dependent magnetic field along the atomic beam compensates the Doppler shift with the Zeeman shift.

An detailed description of our Zeeman slower design can be found in (Taglieber, 2008).

3.4 Laser systems

Cold atom experiments require special laser systems for trapping, detecting and manipulation. Working with three different species simultaneously is a challenging aspect on the laser systems. In the final stage of the experiment with Feshbach resonances, 4 laser systems with 14 lasers and 8 laser locks are used. All of them have to stay in lock for several hours and laser power stability is a crucial aspect to guarantee constant atom numbers. For the build-up special care is taken on stability and on reliability. All laser systems are continuously improved during handling the experiment.

In the following, I will first describe the energy levels and then the laser systems of the three species. Finally, I show the setup for optically trapping the atoms in a far detuned optical dipole trap (ODT).

3.4.1 Energy levels

This section briefly describes specific properties of ^{87}Rb, ^{40}K and ^{6}Li, the relevant energy levels and the corresponding transitions for trapping, detecting and manipulation.

The well-known ^{87}Rb atom is magneto-optically trapped and laser cooled on the D_2-line $|2S_{1/2}, F = 2\rangle \leftrightarrow |5P_{3/2}, F' = 3\rangle$ with a wavelength of 780 nm. The same transition is also used for detecting the atoms by absorption imaging. A repumper, acting on $|5S_{1/2}, F = 1\rangle \rightarrow |5P_{3/2}, F' = 2\rangle$, brings the atoms back into the cycling transition. Optical pumping (OP) is done on the $|5S_{1/2}, F = 2\rangle \leftrightarrow |5P_{3/2}, F' = 2\rangle$ transition.

For the fermionic isotope ^{40}K the trapping transition is between the lowest lying levels $|4S_{1/2}, F = 9/2\rangle \leftrightarrow |4P_{3/2}, F' = 11/2\rangle$ of the D_2-line (767 nm). The repumper transition is $|4S_{1/2}, F = 7/2\rangle \leftrightarrow |4P_{3/2}, F' = 9/2\rangle$ and requires higher laser intensity than in the rubidium case due to the comparatively smaller branching ratio between the excited states. Imaging can be done on closed transitions at variable magnetic fields. A special issue of ^{40}K is the inverted hyperfine structure due to a positive nuclear g-factor. For stability arguments in atomic mixtures this is an important aspect (see Sec. 5.1.2).

In ^{6}Li the hyperfine level structure of the excited $|2P_{1/2}\rangle$ state is on the order of the linewidth and thus not optically resolvable. In the MOT case, lithium can be treated therefore as an effective three level system between the two hyperfine groundstates of $|2S_{1/2}\rangle$ and the excited $|2P_{1/2}\rangle$ state. As a consequence, the pumping transition $|2S_{1/2}, F = 3/2\rangle \leftrightarrow |2P_{3/2}, F' = 5/2\rangle$ and repumper transition $|2S_{1/2}, F = 1/2\rangle \leftrightarrow |2P_{3/2}, F' = 1/2, 3/2\rangle$ on the D_2-line (671 nm) have comparable population rates and both of them need balanced laser intensities for MOT operation. Another important result is the inefficiency of polarization gradient cooling. Moreover, to have a dark state, optical pumping is carried out on the D_1-line $|2S_{1/2}, F = 3/2\rangle \leftrightarrow |2P_{1/2}, F' = 3/2\rangle$ and not on the D_2-line. Imaging, as in the potassium case, is possible at variable magnetic fields. For low field imaging, additional repumping light pumps the atoms back into the cycling transition.

3.4.2 Rubidium

The rubidium system is based on three external cavity diode lasers (ECDL) (Wieman and Hollberg, 1991; Ricci et al., 1995) that are frequency stabilized by either saturated absorption frequency-modulation (FM) spectroscopy (Bjorklund et al., 1983; Drever et al., 1983), by a

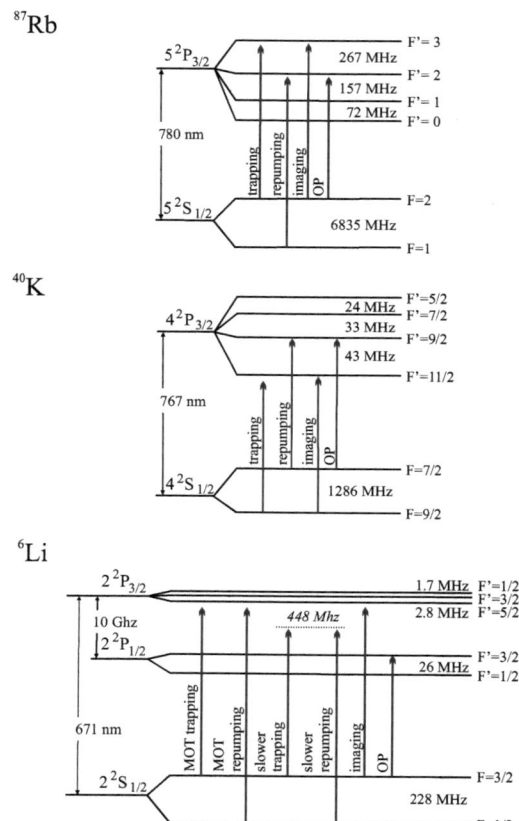

Figure 3.2: Atomic energies levels and transitions that are used to trap and detect rubidium, potassium and lithium.

3.4 Laser systems

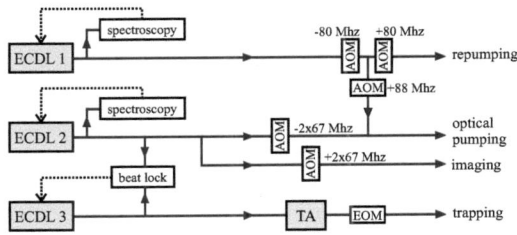

Figure 3.3: Schematic design of the ^{87}Rb laser system.

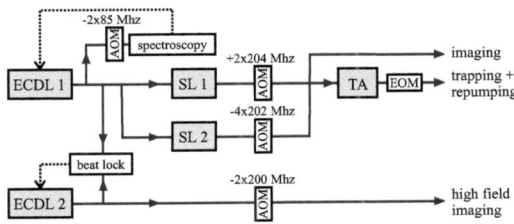

Figure 3.4: Schematic design of the ^{40}K laser system.

Doppler-free dichroic lock (DFDL) (Wasik *et al.*, 2002) or by a beat lock technique (Schünemann *et al.*, 1999).

Laser ECDL1 provides the necessary repumper light for trapping and optical pumping. Laser ECDL2 gives the light for imaging and optical pumping. Laser light from ECDL3 is amplified by an commercial tapered amplifier (TA) system (Toptica, TA-780) and gives 280 mW of trapping light after an optical fiber. The light intensity is controlled by an electro-optic modulator (EOM) (Gsänger, LM 0202). Acousto-optic modulators (AOM) are used to shift laser light frequency in the desired way and to switch off the light that goes into optical fibers for cleaning. Residual light is blocked by mechanical shutters. Further details can be found in (Taglieber, 2008).

3.4.3 Potassium

The laser system for potassium uses two ECDL and two slave lasers (SL). These lasers are composed of AR-coated laser diodes (Eagleyard, EYP-RWE-0790) with high output powers (ECDL 20 mW, SL 40 mW). The coating allows to tune the wavelength easily from the free-running wavelength at 790 nm to 767 nm.

Due to the small natural abundance of fermionic potassium, ECDL1 is frequency stabilized by an FM lock on the cross-over signal between $|4S_{1/2}, F = 1\rangle \rightarrow |4P_{3/2}, F' = 0, 1, 2\rangle$ and $|4S_{1/2}, F = 2\rangle \rightarrow |4P_{3/2}, F' = 1, 2, 3\rangle$ of the D_2-line in ^{39}K. To improve the lock stability, the spectroscopy cell is temperature stabilized additionally to 45.0 °C. This increases the error signal of the locking signal by more than a factor of three.

31

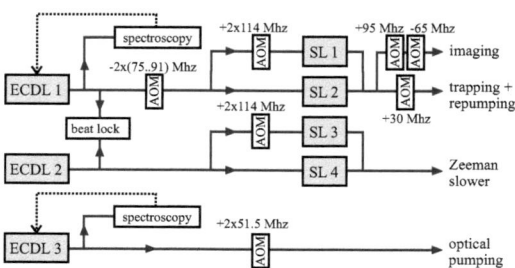

Figure 3.5: Schematic design of the ^{6}Li laser system.

Laser light from ECDL1 is divided in two paths that seeds two slave lasers SL1 and SL2 for light amplification. A double and a quadruple pass AOM controls the frequency. Both beams are then co-aligned on a beam splitter. One beam is then coupled into a fiber for optical pumping and imaging. The other one is amplified by an self-made TA booster system (TA chip by Eagleyard, EYP-TPA-0765-01500-3006-CMT03) for trapping and repumping. After an optical fiber for cleaning the spectral mode, typically 350 mW are achieved for the trapping transition.

Finally, an extra ECDL2 is referenced on ECDL1 by an offset lock and provides light for high-field imaging. The offset frequency is given by an computer controlled signal generator (R&S SML 02).

3.4.4 Lithium

A first version of the lithium laser system can be found in my diploma thesis (Voigt, 2004) and is significantly improved and extended for the measurements discussed in Cha. 5 and 6.

The old laser diodes (Panasonic, LNCQ 05 PS), which are used for the measurements discussed in Cha. 4, are exchanged by laser diodes with a much higher laser output power (Mitsubishi, ML101J27). They are specified with an output power of 120 mW and a free-running wavelength of 661 nm at 25 °C. To run them at 671 nm the laser diodes are heated up to more than 70 °C and the currents are 250 mA for the ECDL and 310 mA for the SL, well above specification. A test laser, ran continuously with these settings for more than half a year, showed no degradation. For the slave lasers stable seeding has been obtained for seeding powers between 100 µW and a few mW. Typical output powers for the slave lasers after the isolator are 80 mW.

The larger available laser power also allows to change some optical paths. The frequency lock of ECDL1 is now optically decoupled from the first AOM to allow fast frequency changes. Moreover, an extra AOM is added in front of the trapping and repumper fiber to control the laser light intensity. These two last changes are important for a reliable operation of the lithium compressed MOT stage (Petrich et al., 1994) of the experimental cycle (see Sec. 5.1.1).

With the larger trapping and slower power the atom number of the Li-MOT is increased by more than two orders of magnitude than in the previous case.

For optical pumping it is beneficial to optically pump the atoms into a dark state. The only transition with an optically resolvable dark state is the D_1-line. Therefore, a third ECDL3, running on the D_1-line, is installed in the existing setup. The frequency is stabilized with an FM-lock using the same spectroscopy cell as ECDL1.

3.4.5 Optical system for MOT, detection and optical pumping

The vacuum chamber and the laser systems, described above, are installed on two different optical tables. For MOT operation, optical pumping and detection laser light is coupled into polarization maintaining optical fibers to guide the light to the desired position. The fibers also clean the spatial profile of the beams and increases the beam quality, which is important for a robust MOT and for high quality absorption images.

Simultaneous trapping of all three species in a MOT requires to superimpose the individual laser beams for trapping and repumping on six MOT beams, four in the horizontal and two in the vertical plane. In contrast to lithium and potassium, repumping light for rubidium is only in the horizontal beams. The trapping power for ^{87}Rb and ^{40}K is equal in all six beams, respectively. Trapping and repumper light intensity for ^{6}Li is comparable within each beam and the vertical beams have two times more power than the horizontal ones. A detailed description of the optical setup is given in (Taglieber, 2008).

3.4.6 Optical dipole trap

For the exploration of Feshbach resonances the atoms are loaded into a far-off resonant optical dipole trap (ODT). This allows trapping of different magnetic spin states that are not magnetically trappable. In addition, trap frequencies and geometry are easily controlled by light intensity and beam waist. Furthermore, arbitrary homogeneous magnetic fields can be applied without affecting the trap frequencies. For a review article on optical traps I refer to (Grimm et al., 2000).

In this chapter I briefly describe the theory of an ODT and show the realized optical setup used within this thesis.

3.4.6.1 Principle

The interaction of an atom with a laser field is modeled by considering the atom as a oscillator driven by a classical radiation field. The electric field $\mathbf{E} = \mathbf{E}_0 \cos(\omega t)$ of the laser light induces a dipole moment

$$\mathbf{p} = \alpha(\omega)\, \mathbf{E} \tag{3.2}$$

with the complex polarizability $\alpha(\omega)$. $\mathbf{p}$ oscillates with the driving frequency ω.

The conservative dipole interaction potential can then be calculated to

$$U_{\text{dip}} = -\frac{1}{2}\langle \mathbf{p}\cdot\mathbf{E}\rangle = -\frac{1}{2\epsilon_0 c}\,\text{Re}(\alpha)\,I\,. \tag{3.3}$$

Here, the potential is proportional to the light intensity $I = 2\epsilon_0 c |\mathbf{E}|^2$. The brackets $\langle\cdot\rangle$ denote time averaging over rapid oscillating terms.

3. Experimental Setup

In addition, the scattering rate by absorption and spontaneous re-emission of photons is related to the imaginary part of the polarizability through

$$\Gamma_{\text{sc}} = \frac{1}{\hbar\,\epsilon_0\,c}\,\text{Im}(\alpha)\,I\,. \quad (3.4)$$

For a far-detuned laser field in the low saturation limit a semi-classical approximation for the polarizability can be used to calculate the resulting dipole potential and scattering rate (Grimm et al., 2000)

$$U_{\text{dip}}(\mathbf{r}) = -\frac{3\,\pi\,c^2}{2\,\omega_0^3}\left(\frac{\Gamma}{\omega_0-\omega}+\frac{\Gamma}{\omega_0+\omega}\right) I(\mathbf{r}) \quad (3.5)$$

$$\Gamma_{\text{sc}}(\mathbf{r}) = -\frac{K}{\hbar}\,U_{\text{dip}}(\mathbf{r})\,,\ \text{with}\ K = \left(\frac{\omega}{\omega_0}\right)^3\left(\frac{\Gamma}{\omega_0-\omega}+\frac{\Gamma}{\omega_0+\omega}\right). \quad (3.6)$$

In these formulae ω_0 is the resonance frequency and Γ is the spontaneous decay rate of the excited state. The detuning $\Delta = \omega - \omega_0$ determines whether the resulting potential is attractive $\Delta < 0$ ("red" detuned) or repulsive $\Delta > 0$ ("blue" detuned). In the attractive case, the induced dipole moment oscillates in phase with the electric driving field of the laser and the atom is attracted to the light intensity maxima. In the repulsive case, on the other hand, the dipole moment oscillates out of phase with the electric driving field and the atom is repelled from the laser light. For detunings $|\Delta| \ll \omega_0$ the dipole potential scales as I/Δ and the scattering rate as I/Δ^2. To reduce scattering and heating of atoms, large detunings are thus favorable.

The heating rate of the atoms can be calculated through

$$\dot{T} = \frac{2/3}{1+\kappa}\,T_{\text{rec}}\,\overline{\Gamma}_{\text{sc}}\,,\ \text{with}\ \overline{\Gamma}_{\text{sc}} = -\frac{K}{\hbar}\left(U_0 + \frac{3}{2}k_{\text{B}}\,T\right). \quad (3.7)$$

In this expression κ is a trap specific parameter, that is $\kappa = 1$ for a 3D harmonic trap (for details see Grimm et al. (2000)). $\overline{\Gamma}_{\text{sc}}$ is the average scattering rate for a cloud of temperature T and $U_0 = U_{\text{dip}}(0)$ is the trap depth.

Atoms can be trapped in the focus of a red detuned Gaussian laser beam. In the harmonic approximation and neglecting gravity, the angular trapping frequencies at the trap center $\mathbf{r} = 0$ are

$$\omega_j = \sqrt{\frac{4\,U_0}{m_A\,w_{0j}}} \quad (3.8)$$

$$\omega_z = \sqrt{\frac{2\,U_0}{m_A\,z_r}}. \quad (3.9)$$

Here ω_j ($j = x, y$) are the radial frequencies and ω_z is the axial frequency along the propagation direction of the laser beam. The beam waist in the focus is w_{0j} and the corresponding Rayleigh length is $z_r = \pi\,w_{0x}\,w_{0y}/\lambda$ with the laser wavelength λ. $U_0 = |U_{\text{dip}}(\mathbf{r}=0)|$ is the trap depth in the center.

3.4.6.2 Technical realization

In the framework of this thesis, an ODT of two perpendicularly crossing laser beams is realized. This offers high atomic densities and a small spatial extension of the cloud, which is essential

3.4 Laser systems

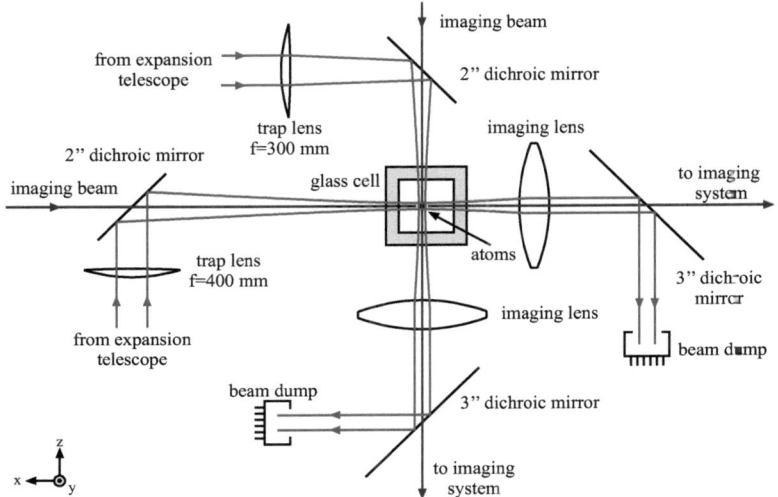

Figure 3.6: Setup of the optical dipole trap. Two beams in crossed configuration along the x- and z-direction are focused down to the atom position in the glass cell. Dichroic mirrors overlap the ODT beams with the imaging light.

for a high precision magnetic field control. The crossed ODT is formed by one beam in the horizontal plane, along x-direction, and the other beam in the vertical plane, along z-direction. Compared to two beams in the horizontal plane, this choice is advantageous for evaporative cooling in the ODT. In the horizontal-vertical setup the contribution of the vertical beam to the trapping frequency and trapping depth along z-direction is typically negligible compared to the contribution of the horizontal beam. Thus, for evaporative cooling the vertical beam power can be kept high, while lowering the horizontal beam power and consequently lowering the trapping depth along z-direction. This accelerates the evaporation sequence, since the resulting trapping frequencies and atomic densities are higher in this case than in a horizontal-horizontal setup. The schematics of the optical dipole trap is illustrated in Fig. 3.6.

The laser source is a ytterbium based single-mode, single-frequency fiber laser (IPG photonics YLR-20-LP-SF) with a maximum output power of 20 W and a laser wavelength of 1064 nm. The output beam of the fiber laser is divided into two beams. AOMs shift each of them by plus or minus 110 MHz in frequency to avoid interferences in the crossed ODT. The two beams are then coupled into optical fibers for cleaning. After the fiber, the beams are expanded and the polarization is selected by beam splitters. Low-reflective beam splitters send a small part of the beams on photodiodes for controlling the beam intensities. A detailed description on the PI-regulation circuits is given in the diploma thesis (Wieser, 2006). Electromagnetic shielding of the electronics reduces radio frequency interference rectification errors and ensures sensitive control, even in the presence of high power RF or MW signals (see Sec. 3.9). At the atom position in the glass cell, the beams are focused down to a $1/e^2$ radius of $w_{\text{hor}} = 55\,\mu\text{m}$ and

35

3. Experimental Setup

$w_\text{ver} = 50\,\mu\text{m}$ for the horizontal and vertical beam, respectively. Special designed dichroic mirrors (at a 45° angle of incidence: $R > 99.8\,\%$ for $1064\,\text{nm}$, $T = 90 - 94\,\%$ for $670 - 780\,\text{nm}$) overlaps the ODT beams with the imaging light and send them into beam dumps after the glass cell. The large diameters of the dichroic mirrors allow high resolution imaging with a large numerical aperture.

3.5 Absorption imaging

Absorption imaging is a fundamental technique to extract relevant information from ultracold gases, e. g. atom number, temperature and density profile. The technique is widely discussed in literature and I refer to the following overview article (Ketterle et al., 1999). For absorption imaging near-resonant photons are sent onto atoms and scattered into a solid angle of 4π. A lens system images the resulting shadow cast by the atoms onto a charge-coupled device (CCD) camera. Postprocessing the data with noise correlation interferometry (Altman et al., 2004) even allows to reveal information on strongly correlated states (Greiner et al., 2005; Fölling et al., 2005).

3.5.1 Principle

The recorded shadow cast represents a two-dimensional intensity distribution in the x-y plane. For simplicity, imaging is assumed along the z-direction. The optical density of the atomic cloud $OD(x, y)$ is given by the transmission $T(x, y)$, which is evaluated by taking three images. The first one is the absorption image $I_\text{abs}(x, y)$ with atoms in the absorption beam. The second one, the reference image $I_\text{ref}(x, y)$, is taken with the same imaging parameters but without any atoms. The last one is a background image $I_\text{bg}(x, y)$ with no beam light on the camera and it represents the information of residual stray light and dark counts of the CCD. The optical density and the transmission have then the following expression

$$OD(x, y) = -\ln T(x, y)\,,\ \text{with}\, T(x, y) = \frac{I_\text{abs}(x, y) - I_\text{bg}(x, y)}{I_\text{ref}(x, y) - I_\text{bg}(x, y)}\,. \tag{3.10}$$

Evaluating the transmission through three different images significantly improves the image quality, since unknown parameters in the optical setup are canceled out, e.g. interference effects and beam imperfections.

In the limit of small saturation intensity I_s, i. e. $S \ll 1 + (2\,\delta/\Gamma)^2$ with $S = I/I_\text{s}$, the beam intensity I, the transition linewidth Γ and optical detuning δ, Lambert-Beers's law is applicable

$$OD(x, y) = \sigma_\text{ph}\,\tilde{n}\,(x, y)\,. \tag{3.11}$$

Here the column density $\tilde{n}(x, y) = \int n(x, y, z)\,dz$ is the integration of the atomic density $n(x, y, z)$ along z-direction and σ_ph is the absorption cross section. For a two-level system with the transition wavelength λ_0 the cross section σ_ph has the simple form

$$\sigma_\text{ph} = \frac{3\,\lambda_0^2}{2\,\pi}\frac{1}{1+(2\,\delta/\Gamma)^2}\,. \tag{3.12}$$

3.5 Absorption imaging

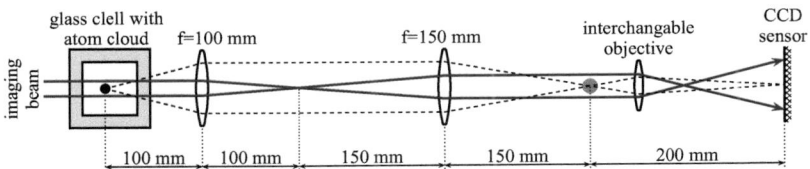

Figure 3.7: Optical setup of the imaging system along one direction. Imaging along all three axes with the same camera position is achieved by appropriate folding the beam and inserting elliptical gold mirrors.

3.5.2 Optical setup

The optical setup (see Fig. 3.7) is based on three achromatic lenses with broadband AR coating that image the cloud on a CCD sensor. The first two lenses form a Rayleigh telescope that creates an image of the cloud in an intermediate plane. The third lens is an objective that creates the image on the CCD sensor. In the experiment, imaging along all three axes with the same camera position is possible. This is achieved by using three Rayleigh telescopes in each spatial direction, and each of them can be addressed by the same objective by inserting steering mirrors in a reproducible way.

Large diameters of the lenses (2 inches for the first two and 1 inch for the last lens) and close placement of the first lens to the atom position (100 mm) ensure high resolution of the imaging system. The measured resolution is about 7 µm. The last lens in front of the CCD camera is exchangeable. This allows to change the optical magnification by keeping the total length of the optical system constant and not changing the camera position in the focus position. Possible magnifications used within this work range from $M = 0.3$ to $M = 7$. The magnification is either determined with a high-resolution test pattern (Edmund Optics, USAF res target, NT 38-257) or by releasing an atomic sample from the trap and measuring the falling distance due to gravity.

Two different cameras are used. The first camera (Apogee, AP1E) has 768×512 pixels and a pixel size of $9\,\mu\text{m} \times 9\,\mu\text{m}$ on the CCD-sensor (Kodak, KAF-0401E).

The second camera, that was later used for the experiments with Feshbach molecules (see Cha. 6), is a frame transfer camera (Andor, A-DU934N-BRD) with a back-illuminated CCD sensor for higher sensitivity. This sensor has 1024×1024 pixels and a pixel size of $13\,\mu\text{m} \times 13\,\mu\text{m}$. The so-called kinematics mode of this camera allows imaging with high repetition rates. For two subsequent images we typically cover half of the sensor with a mask to not illuminate this part during the imaging pulse. After the pulse the illuminated part is shifted below the covered one and a second image can be taken. The used line shift of this camera is $2.975\,\mu\text{s/lines}$ and allows to take two images of 1024×512 pixels with a delay of 2 ms. This delay is by far much shorter than the time needed to transfer the image to the computer and reinitialization for the next image, which is typically 800 ms and thus too long to image reliable an atomic cloud. Imaging twice or more within one run shortens the overall data acquisition time by at least a factor of two and significantly reduces fluctuations.

Backreflections and vibrations of the optics affect the image quality by unwanted interferences. Therefore the optics are well chosen and mounted. Nevertheless, the image quality is affected by a small interference effect coming from surface reflections at the camera sensor and the glass cell. This interference is significantly reduced by a nice trick with a $\lambda/4$ wave plate that is inserted in front of the last lens at the camera.

3.5.3 State selective imaging

State selective imaging allows to image different atomic or molecular states independently. In this section two different techniques are presented that are used in Cha. 5 and 6.

The first one is separation of two spin states by applying a Stern-Gerlach (SG) force during time-of-flight before the image is being taken. Here a magnetic field gradient separates states of different magnetic moments. A separation of different atomic states normally requires a low magnetic offset field to work in the Zeeman and not in the Stark field regime. Furthermore, high magnetic field gradients are necessary for fast separation and small expansion times of the clouds. Thus, two magnetic coils, mounted very close to the glass cell, are used for the SG separation. The same coils are also used for the magnetic transfer and the QUIC trap (see below Sec. 3.6), but are now operated in anti-Helmholtz configuration. This setup allows to achieve low magnetic offset fields with high magnetic field gradients. The spatial separation is in the $x-z$ plane.

The second state selective detection technique is imaging at high magnetic fields, where the energy separation of the atomic states is much larger than the transition linewidth Γ. Since the magnetic field is kept high, this technique even allows to explore the interaction dynamics at various magnetic fields and scattering lengths. In the experiment, the imaging direction is along the y-direction, i.e. perpendicular to the magnetic field **B**. The imaging light is linearly polarized perpendicular to **B**. Therefore, the effective absorption cross section is cut into halves for a circular polarized transition. In general, the cross section σ_{ph} is a function of the magnetic field and of the involved transition. The cross sections for the optical transitions at high magnetic fields, which are used in Cha. 6, are calibrated with measurements at zero magnetic field. The calibration is consistent with a theoretical calculation according to (Windholz and Musso, 1988).

3.6 Magnetic trapping

In addition to optical trapping, atoms can be trapped purely magnetically if they have a permanent magnetic dipole moment.

In the following I explain the principle of magnetic trapping and give an overview of the magnetic transport that moves the atomic clouds from the MOT chamber into the UHV chamber. Here the clouds are transferred into a QUIC trap for evaporative and sympathetic cooling.

3.6.1 Principle

The potential energy of an atom in an external magnetic field $\mathbf{B}(\mathbf{r})$ is related to

$$E(\mathbf{r}) = -\boldsymbol{\mu} \cdot \mathbf{B}(\mathbf{r}). \tag{3.13}$$

Here $\boldsymbol{\mu}$ is the magnetic dipole moment of the atomic state. In the linear Zeeman regime, i.e. for low enough magnetic fields, the energy can be expressed as $E = m_\text{F}\, g_\text{F}\, \mu_\text{B}\, |\mathbf{B}(\mathbf{r})|$, where g_F is the Landé g-factor of the hyperfine state and m_F is the quantum number of the corresponding state. The states that fulfills the condition $m_F\, g_F > 0$ are called "low field seekers" and are magnetically trappable. In contrast to this, the "high field seekers" (states with $m_F\, g_F < 0$) are not magnetically trappable. This is due to Maxwell's equations, that do not allow a field maximum in free space (Wing, 1984).

A widely-used magnetic trap is a quadrupole trap, which has a zero crossing in the trap center. According to Maxwell's equations, the gradients fulfill the general condition $div\mathbf{B} = \frac{\partial B_x}{\partial x} + \frac{\partial B_y}{\partial y} + \frac{\partial B_z}{\partial z} = 0$. Therefore, the quadrupole trap can be characterized by only two parameters, e.g. the gradient $\frac{\partial B_z}{\partial z}$ and the aspect ratio $A = \frac{\partial B_y}{\partial y} / \frac{\partial B_x}{\partial x}$. The most simplest way to realize such a trap is a spherical quadrupole trap with $A = 1$, carried out by two magnetic coils in anti-Helmholtz configuration (axis of the coils are aligned along the z-direction).

To stay trapped, the magnetic moment of the atoms has to adiabatically follow the local magnetic field vector. The adiabaticity criterion is given by

$$\left| (\mathbf{v} \cdot \nabla) \frac{\mathbf{B}}{|\mathbf{B}|} \right| \ll \omega_L . \qquad (3.14)$$

Here $\mathbf{v}$ is the atom velocity and $\omega_L = \mu\, B/\hbar$ is the Larmor frequency. For non-adiabatic motion the atom can make spin flips to untrapped states, so called Majorana spin flips (Majorana, 1932).

In magnetic traps these losses take place in a region with a "hole" size of $\sqrt{v\, \hbar/\mu\, B'}$ at the trap center (B' is the radial gradient) and the lifetime scales as $m\, T^2/B'^2$ (m is the atom mass, T is the temperature of the gas) (Petrich et al., 1995). In a quadrupole trap with typical temperatures of 100 µK and gradients of up to 100 G/cm the Majorana losses can be neglected for our three species. The losses become significant at lower temperatures. For evaporative and sympathetic cooling therefore, the atoms are loaded into a QUIC trap, see Sec. 3.6 3 below.

3.6.2 Magnetic transport

All three atomic species are loaded simultaneously in a magneto-optical trap (MOT) in the MOT chamber. To increase the lifetime of the gas the atoms are transferred into the UHV chamber. The transfer, which is illustrated in Fig. 3.8, is realized by the magnetic transfer technique with a 90° corner (Greiner et al., 2001). This allows excellent optical access in all three directions, because no second MOT optics are required. For a detailed description of the transfer I refer the reader to (Taglieber, 2008).

The magnetic transfer has a total transfer distance of 39 cm and is based on 13 overlapping quadrupole coil pairs and a push coil at the beginning of the transfer process. The current of the coil pairs is ramped up and down from one to the other in such a way that the atoms, trapped at the minimum magnetic field position, are continuously transferred from the MOT to the UHV chamber. To avoid heating, the acceleration and the variation of the trap geometry are optimized. The gradient along the z-direction $\frac{\partial B_z}{\partial z}$ is kept constant and the aspect ratio is changed continuously from $A = 1$ at the beginning to $A = 1.662$ during the transport. The ratio is changed again to $A = 1$ at the corner and at the end of the transfer. The value of $A = 1.662$ corresponds to the aspect ratio of a cloud sitting between two coil pairs. During transport, this value is achieved by controlling the current through three adjacent coil pairs. The power supply (Lambda, ESS 30-500) is run in constant voltage (CV) mode and MOSFET banks control the current through each coil pair.

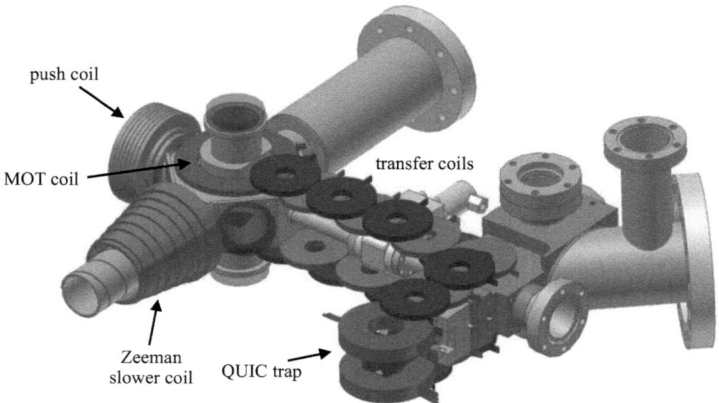

Figure 3.8: After simultaneously loading the three species in a MOT, the atoms are transferred from the MOT chamber into the UHV chamber by a magnetic transport with a 90° corner.

3.6.3 QUIC trap

Due to Majarona losses (as described above in Sec. 3.6.1), cooling to quantum degeneracy does not work in a simple quadrupole trap. Several solutions exist to overcome this problem in magnetic traps. The first BEC (Anderson et al., 1995) was realized in a time-averaged, orbiting potential (TOP) (Petrich et al., 1995). Another solution is to "plug" the "hole" in a quadrupole trap by a repulsive potential with a blue-detuned laser (Davis et al., 1995). A common magnetic trap is the Ioffe-Pritchard trap (Pritchard, 1983) with a non-zero magnetic field minimum. For the work presented here, a quadrupole-Ioffe-configuration (QUIC) trap is used (Esslinger et al., 1998). It offers a simple geometry and a low power consumption.

The QUIC trap is based on a pair of quadrupole coils in anti-Helmholtz configuration and a Ioffe coil that generates a magnetic curvature and a finite magnetic offset field B_0 (see Fig. 3.9). In addition two compensation coils produce a homogeneous magnetic field. Bypassing these coils allows to adjust B_0. The loading sequence of the atoms into the QUIC trap is optimized with the Feshbach coils.

The QUIC trap can be characterized by three parameters: B_0, the axial gradient 2α of the quadrupole field and the curvature β along the trap axis. For a total current of 30 A, which is used within this thesis, $\alpha = 146.1(7)\,\mathrm{G/cm}$ and $\beta = 254.0(5)\,\mathrm{G/cm^2}$ (see Sec. 5.1.4). In the harmonic approximation, i. e. $k_\mathrm{B} T \ll \mu_\mathrm{B}\, g_\mathrm{F}\, m_\mathrm{F}\, B_0$, the axial and radial trapping frequencies (ω_x, ω_ρ) of the QUIC are given then by

$$\omega_x = \sqrt{\frac{\mu_\mathrm{B}\, g_\mathrm{F}\, m_\mathrm{F}}{m} \beta} \tag{3.15}$$

$$\omega_\rho = \sqrt{\frac{\mu_\mathrm{B}\, g_\mathrm{F}\, m_\mathrm{F}}{m} \left(\frac{9\,\alpha^2}{4\,B_0} - \frac{\beta}{2} \right)}. \tag{3.16}$$

3.7 Feshbach coils

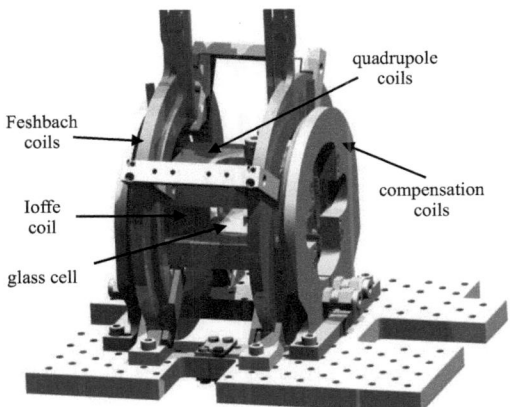

Figure 3.9: The quadrupole-Ioffe-configuration (QUIC) trap is based on a pair of quadrupole coils in anti-Helmholtz configuration and a Ioffe coil that generates a magnetic curvature and a finite magnetic offset field. Additional compensation coils produce a homogeneous magnetic field. The Feshbach coils are placed close to the glass cell. Special mounting of the coils guarantees stable operation.

For stability reasons, the current flow through the QUIC coils is arranged in series to minimize fluctuations due to current noise of the power supply (Delta Elektronika, SM45-70D, max. 53 V/ 70 A). To further increase the stability and reproducibility of the trap, the QUIC and Feshbach coils are mechanically decoupled from the magnetic transport system. The QUIC trap can be switched-off within 400 µs by the help of a varistor.

3.7 Feshbach coils

The exploration of atomic gases in the vicinity of Feshbach resonances requires an ultra stable, homogeneous magnetic field with high precision control. At the build-up time, no Feshbach resonances between ^{6}Li and ^{40}K and between ^{6}Li and ^{87}Rb were known. In the following, I present a versatile magnetic Feshbach design with a high precision control.

Feshbach coils The coils for the magnetic Feshbach fields has to fulfill high mechanical stability and a small inductivity for fast field modulations.

The magnetic coils are made of a rectangular, hollow copper wire with the outer dimensions of $4 \times 4\,\text{mm}^2$ and an inner diameter of 2.5 mm (Wolverine Tube Europe BV, The Netherlands). The wire is electrically insulated with fiberglass filaments (SW Wire Co., USA). Each coil consists of two parts, one with one layer of 3 windings and the other one with two layers of 7 windings. The outer diameters are 15.4 cm and 18.6 cm, respectively. In both cases the inner diameters are 12.8 cm. The wire is glued with a special high heat conductance epoxy. Additional bars between the coils guarantee an increased mechanical stability. The hollow wire

3. Experimental Setup

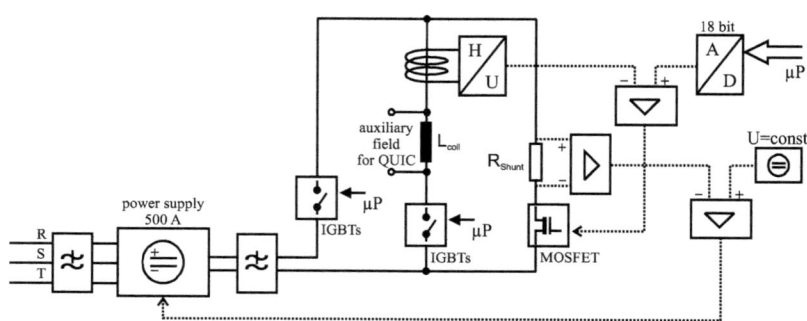

Figure 3.10: Current stabilization diagram for the magnetic Feshbach fields. The current through the Feshbach coils is controlled by two feedback loops: a fast regulation through a current by-pass parallel to the Feshbach coils and a slow regulation of the power supply.

allows excellent water cooling. The Feshbach coils are placed between the compensation coils of the QUIC trap (see Fig. 3.9) and do not affect the optical access to the cell.

The magnetic field of the Feshbach coils is calibrated with known hyperfine or Zeeman transitions (see Sec. 6.2). The field scales as 1.9 G/A. The power supply is a second one from Lambda (ESS 30-500), the same model as already used for the magnetic transfer. Both of them can be operated in master-slave mode with a maximum current of 1000 A and a maximum field of about 1800 G.

The measured inductivity of the coils is only 137 µH and allows fast modulation of the magnetic field. Fast switch-off times of the field is facilitated by varistors. Typical initial slopes are 820 G/cm. For typical magnetic field values of 150 G the field curvature at the center of the coils is 1.19 G/cm². Due to technical constraints, the atomic position is slightly shifted with respect to the coil center. The field gradient and curvature at the atom position are 385 mG/cm and 1.12 G/cm², respectively. For typical cloud sizes of < 10 µm in the crossed ODT the magnetic field variation over the cloud is smaller than 0.4 mG.

Stabilization circuit The basic idea to control the Feshbach current is a fast controlled current by-pass parallel to the Feshbach coils combined with a slow feedback control of the current of the power supply (see Fig. 3.10).

The current through the Feshbach coils is measured by a current transducer (Danfysik, Ultrastab 867-1000I HF) that is based on the balanced zero flux detection principle. This device produces a current that is proportional to the Feshbach current. The transducer has a high reproducibility and the temperature drifts are specified to be 0.5 ppm/K. Special burden resistors (< 1 ppm/K) convert the current into a voltage for a set-actual comparison. The reference value is given by a 18 bit digital-to-analog converter (DAC) (Analog Devices, AD760). A microprocessor controls the DAC value in real-time. It is possible to run arbitrary wave forms. A PI control gives the feedback signal on a MOSFET (IXYS, IXFN 230N10) in the current by-pass.

To keep the average by-pass current at 10 A the by-pass current is first measured by a shunt resistor, and then another PI control regulates the power supply. This two-folded control enables to control the Feshbach current with only one value over the whole current range. Furthermore, compared to the in general slow power supply control, the Feshbach current can be controlled with a much higher bandwidth. In addition, the constant by-pass current sets the working point of the MOSFET at an almost constant position and thus allows working at different magnetic fields without varying the control parameters. An alternative solution of the current control is a single feedback loop with a MOSFET in series with the Feshbach coils. This solution would require to vary the control parameters and to vary the voltage of the power supply manually to minimize power dissipation.

In order to achieve a high reproducibility, the whole electronic control circuit is by itself temperature stabilized to be better than 0.1 °C. Furthermore, particular care is taken on the choice of electronic components. Special line filters suppress noise from the Lambda power supply.

The current can be switched-off by IGBTs (Semikron, SKM 800GA 126D): Two IGBTs in parallel are used for the current through the Feshbach coils and two other IGBTs in parallel switch-off the total current through the by-pass and the Feshbach coils. The last two of them and additional high current relays separate the circuit from the auxiliary magnetic field for the QUIC trap loading procedure.

3.8 Ambient magnetic fields

In addition to the previous magnetic fields, it is important to compensate the earth magnetic field and to characterize magnetic fields noise in the laboratory environment. This is of particular importance for the measurement of the narrow Feshbach resonances as presented in Sec. 6.

3.8.1 Magnetic offset fields

The earth magnetic field at the MOT and UHV chamber is compensated by magnetic field coil pairs in the spatial directions x, y and z. Each coil consists of 20 windings and the current in each coil pair is individually controlled by external power supplies. The current can be switched between two set points via a TTL-signal or it can be arbitrarily controlled via an analog signal. This flexibility allows to add extra guiding fields for imaging or optical pumping.

For the earth magnetic field compensation two different methods are used. At the MOT chamber the ambient magnetic field is zeroed by minimizing the temperature after optical molasses cooling, see also discussion in Sec. 4.2.5.

The magnetic field at the UHV chamber is set to zero by a method employing Hanle spectroscopy. For this purpose, a rubidium cloud of several $10\,\mu\text{K}$ temperature is imaged after 5 ms time-of-flight, so that residual magnetic fields from the QUIC trap have decayed. Imaging of the cloud is performed with sigma light on the $|5S_{1/2}, F=2\rangle \leftrightarrow |5P_{3/2}, F'=2\rangle$ transition with a few hundred photons scattered during the imaging pulse. Additional repumping light on the $|5S_{1/2}, F=1\rangle \rightarrow |5P_{3/2}, F'=2\rangle$ transition brings the atoms back to the $|F=2\rangle$ manifold. If the magnetic field is varied by the compensation coil along the transverse direction to the imaging light propagation, the absorption signal has a minimum at the zero crossing of the magnetic field. The reason is that the atoms are pumped into a dark state. In the other case, when the magnetic field is varied along the longitudinal direction, the absorption signal peaks

3. EXPERIMENTAL SETUP

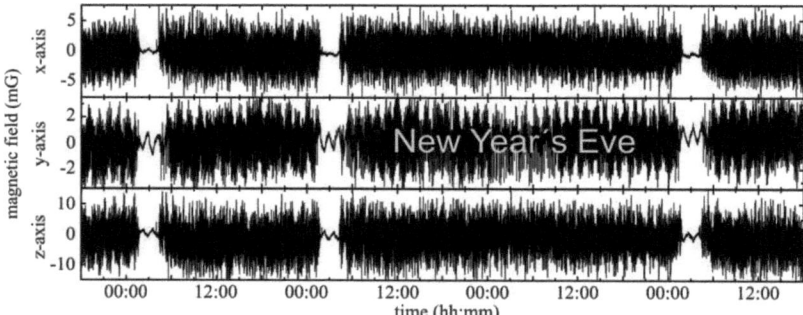

Figure 3.11: External magnetic fields are monitored for the x-, y- and z-direction for several days. The data shows that the magnetic fields drops down in a time-frame of roughly 2:00 a.m. and 4:30 a.m., which is consistent with the timetable of the subway that stops during the night. At New Year's Eve, on the other hand, the subway run the whole night, which is demonstrated by the data.

at the zero crossing, since the dark state is absent. Iterating the method for the different spatial directions minimizes the magnetic field at the trap center.

3.8.2 External magnetic fields

In addition to the constant earth magnetic field, time-dependent magnetic fields from the environment are characterized. There are two time-dependent contributions with different time scales. One is 50 Hz noise and the other one comes from the nearby subway traffic. In the following, the magnetic field is measured with a three-axis, magneto-resistive sensor (Honeywell, HMC 1053).

When all power supplies are switched on, the 50 Hz noise has a peak-to-peak signal of 3.3 mG. In the case, when the power supplies close to the UHV chamber are switched off, the noise is reduced to a peak-to-peak signal of 2.2 mG.

The other magnetic field contribution comes from the subway, that passes by close to the lab. The magnetic field varies slowly on a 20 s time scale. The subway direction is parallel to the Zeeman slower, i.e. along the y-direction in lab coordinates. Whenever a subway runs on the current section, which passes the lab, the magnetic field in the lab changes due to acceleration or deceleration. The dominant magnetic components are in the $x - z$ plane and the magnetic field vector in this plane is consistent with the magnetic field direction generated by the subway. Fig. 3.11 shows the magnetic fields monitored over several days around New Year. The data is taken with a low-pass filter suppressing 50 Hz noise. The largest magnetic field components are along the z-direction and are up to ±10 mG. The magnetic fields drop down in a time-frame of roughly 2:00 a.m. and 4:30 a.m., which is consistent with the timetable of the subway that stops during the night. At New Year's Eve, on the other hand, the subway ran the whole night, which is demonstrated by the data.

For magnetic field sensitive measurements therefore, one is restricted to the small time-frame at night. In order to increase reproducibility, the experimental cycle can be triggered on the 50 Hz signal (see Sec. 3.10).

3.9 RF & MW setup

In this section I present the experimental radio frequency (RF) and microwave (MW) setup that is designed to fulfill different applications. One is evaporative cooling by RF or MW field induced transitions. Another one is state preparation of all three species by an adiabatic rapid passage (ARP) or by an π pulse at different magnetic field strengths. The setup can also be used for molecule RF spectroscopy.

Antennas To cover the necessary frequency bands, different antenna geometries are used. In addition, geometrical constraints have to be considered to keep the excellent optical access high.

RF evaporation typically uses a frequency band of a few 100 kHz to several tens of MHz. This is covered by a quasistatic antenna of two-folded loops with a quadratic shape (26×26 mm^2) of thick copper wire (1 mm). A second antenna of the same design is used for ^{6}Li hyperfine transitions. The application runs from $200 - 300$ MHz (see also Sec. 5.2.2.3 and 6.3) and is therefore frequency matched with a network analyzer. The antenna for ^{40}K hyperfine transitions at ~ 1.3 GHz is folded to a quadratic shape (26×26 mm^2) out of a copper plate (2 mm thickness, 1 cm deepness). This antenna is also frequency matched and for higher surface quality it is chemically etched and tinned (see thesis by (DeMarco, 2001)).

For ^{87}Rb hyperfine transitions at ~ 6.85 GHz different antenna geometries are tested. The finally installed antenna is a $\lambda/2$ dipole antenna ($\lambda/2 \approx 2.2$ cm), which has a comparatively compact design and allows to place it directly at the glass cell close to the atoms without corrupting the excellent optical access.

In order to drive hyperfine transitions, the oscillating magnetic field $\mathbf{B}_{ac}$ of the antenna has to be perpendicular to the local magnetic field $\mathbf{B}_{atom}$ at the atomic position. For high temperatures in the QUIC trap, the dominating magnetic field components are the radial and axial ones. For low temperatures in the QUIC trap and for studies with magnetic Feshbach fields, the dominating term is the axial component along the x-direction. Two positions close to the glass cell come into consideration for the placement of the antennas: One is at the glass side opposite to the Ioffe coil and the other one is at the glass side that is along the magnetic transfer direction. The dipole antenna for ^{87}Rb is placed close to the former one such that $\mathbf{B}_{ac}$ is parallel to the z-direction. The other three antennas are placed close to the latter glass cell side such that $\mathbf{B}_{ac}$ is parallel to the y-direction.

RF and MW sources The RF and MW frequencies are generated by programmable synthesizers and amplified to several watts. Each signal is controlled in amplitude by a voltage-variable attenuator (VVA) (Hittite, HMC346MS8G, DC - 8 GHz, attenuation range 30 dB) and TTL-switches (attenuation > 60 dBm).

In the experiment a versatile programmable RF source is used that is based on a direct digital synthesis (DDS) chip. The system is self-made and the first version is based on a design developed at the AMOLF institute in Amsterdam that uses a chip (Analog Devices, AD9854) with frequencies between $0.2 - 150$ MHz. An improved second version generates frequencies

3. Experimental Setup

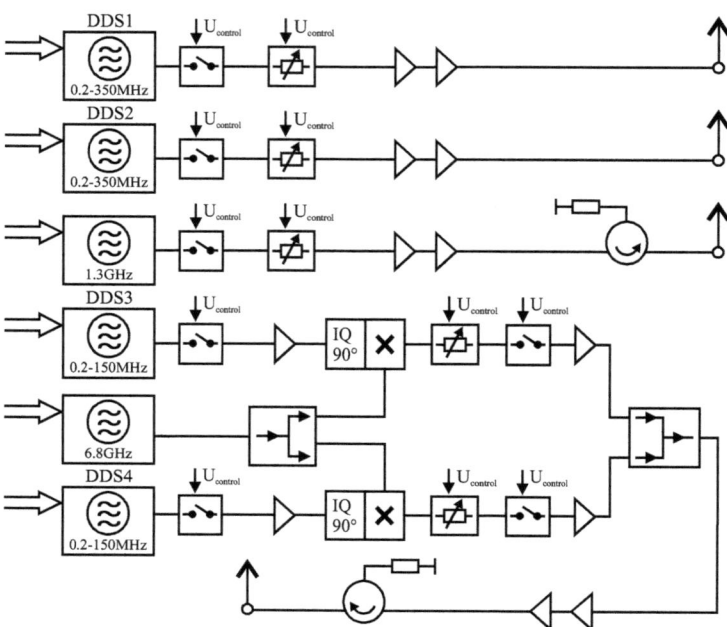

Figure 3.12: The RF & MW layout for Zeeman and hyperfine transitions. The upper two lines are used for RF evaporation and ^{6}Li hyperfine transitions. The third line is for ^{40}K hyperfine transitions. The last three lines are used for evaporative cooling of rubidium by hyperfine transitions.

between 0.2 − 350 MHz (Analog Devices, AD9858). This new DDS version also allows to create phase coherent sweeps as well as phase continuous real-time controlled frequency updates, and thus widens the spectrum of applications.

Two of the DDS boards (DDS1 and DDS2 in Fig. 3.12) generate the frequencies for RF evaporation, state preparation and ^{6}Li hyperfine transitions. The signals can be amplified to a maximum output power of 10 W (HD Communications Corp., HD19169 and Delta RF Technology, LA10-2-512-40), which allows large Rabi frequencies and short transition times (see Sec. 6.3).

For ^{40}K hyperfine transitions the signal, generated by a synthesizer (R&S SML 02, 9 kHz - 2.2 GHz), can be amplified to more than 10 W (Kuhne electronic, KU 2025 A and MKU 1330 A). A circulator protects the main amplifiers from backreflections at the antenna.

In the rubidium MW case, evaporation is performed by driving hyperfine transitions that are state selective and do not affect the other two species (see Sec. 5.2.1.4). In order to clean unwanted rubidium states continuously, a MW system generates two independently controllable MW frequencies. It is a reasonably priced design based on only one MW synthesizer (Systron Donner, 1720, 50 MHz − 18 GHz) and one master amplifier (Kuhne electronic, KU 682 XH-UM, 16 W) (see Fig. 3.12). The MW signal of the synthesizer is split into two parts. Then, each

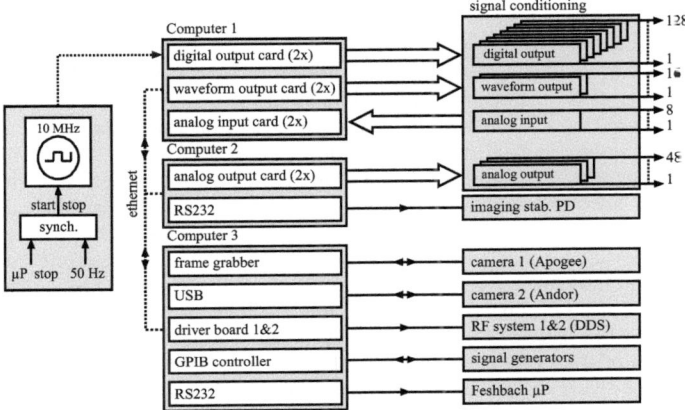

Figure 3.13: Schematic of the experimental control. Three computers control the experimental parameters in real-time. The I/O cards are synchronized by a stable 10 MHz clock source.

MW signal is mixed with an RF signal that is generated by a DDS board (DDS3 and DDS4 in Fig. 3.12). Mixing is performed with single side-band (SSB) mixers that significantly suppress the carrier and the unwanted mirror frequency (typical suppressions are $\sim 40\,\text{dB}$). After passing VVAs and TTL-switches the signals are combined and amplified. Here, special care is taken to work in the linear regime of the MW components and to avoid high-harmonic generations. As before, the circuit is protected by a circulator.

Due to reproducibility reasons, all DDS boards and synthesizers are synchronized to the high-precision 10 MHz reference signal from the Systron Donner synthesizer (aging rate $\leq$ 5 parts $10^{-10}/\text{day}$).

3.10 Experimental Control

Operations with ultracold quantum gases require in general a real-time operation of the experimental parameters. Typically, hundreds of operations are processed sequentially for a single measurement with step sizes between 1 μs and several seconds. To handle such a large amount of parameters conveniently, the operations are computer controlled via three computers, linked through TCP/IP connections. The hard and software design is based on a development at the AMOLF institute in Amsterdam and is described briefly in the following.

The control software is written in LabView and has a graphic interface for convenient programming. New devices and post-processing routines are easily implemented.

Fig. 3.13 shows the current schematic of the experimental control. Different I/O cards are implemented: two digital output cards with a total of 128 channels (Viewpoint Systems, PCI-DIO64), two cards for 16 analog waveform outputs (National Instruments, PCI-6713), one analog input card with 8 channels (National Instruments, PCI6034e), and two cards for 48 analog output channels (United Electronics, PCI-PD2-AO 16/16 and PCI-PD2-AO 32/16).

3. Experimental Setup

These signals are further processed by signal conditioning units, which provide (quasi-)galvanic isolation of the channels and allows to modify each signal individually.

Furthermore, two RS232 interfaces are used to program the flash image stabilization box and the microcontroller of the Feshbach stabilization circuit. In addition, a GPIB controller card allows to address synthesizers and arbitrary waveform generators timely during the experimental cycle. The cameras are controlled by a frame grabber card (Apogee, AP1-E) and by USB, installed on computer 3. The two DDS systems are addressed by two dedicated driver boards.

The I/O cards are synchronized by a common clock source. To further increase the reproducibility, the external 10 MHz reference signal from the Systron Donner synthesizer is used as the clock source for all timed devices of the control system. A digital output can be used to halt the system by interrupting the 10 MHz signal by a gate element. Operation is resumed on the next occurrence of a 50 Hz line trigger signal. In this way the experiment can be efficiently synchronized in spite of the short coherence length of the line signal, which is shorter than the duration of the experimental cycle. This synchronization significantly improved the reproducibility, especially at the final evaporation stages and for RF & MW transitions, which are sensitive to external magnetic field variations (see Sec. 3.8).

Finally, failure protections are implemented, which are especially important for dealing with high current power supplies. The protection system uses the main interlock for all critical devices and is switched-off in case of an alert. Temperatures of the magnetic coils and amplifiers as well as water flow through several channels are monitored and compared with set values with a data acquisition system (Keithley, 2700/E).

Chapter 4

Magneto-optical trapping of three atomic species

In this chapter I present the first magneto-optical trapping of two different fermionic species, ^{6}Li and ^{40}K, and a bosonic species ^{87}Rb. This demonstrates the first trapping of two fermionic species and also of three species in a magneto-optical trap ("triple MOT"). Optimization of the atom numbers in the triple MOT configuration is shown and, in addition, loading and trap losses are characterized. These results paves the way towards a quantum degenerate mixture of two different fermionic species.

Parts of this chapter are published in:
M. Taglieber, A.-C. Voigt, F. Henkel, S. Fray, T. W. Hänsch, and K. Dieckmann, "Simultaneous magneto-optical trapping of three atomic species," Phys. Rev. A **73**, 011402 (2006).

4.1 Experimental setup

The experimental setup for the MOT used within this chapter is shown in Fig. 4.1. As described in Sec. 3.4.5, the beams for pumping and repumping light for all three species are overlapped and aligned in such a way that they build a six-beam MOT of counter-propagating beams with σ^+ and σ^- polarization. The necessary magnetic field is given by a pair of magnetic coils in anti-Helmholtz configuration.

The vacuum system used here, consists of the oven and MOT chamber, as outlined in Sec. 3.2, but without the UHV chamber part. ^{40}K and ^{87}Rb are loaded from atomic vapor dispensers (see Sec. 3.3.1). The potassium dispensers are home-built ones of the first generation with a 3 % enrichment of ^{40}K. Lithium is loaded from a Zeeman slower to guarantee highly efficient loading compared to loading from background gas. The parameters of the Zeeman slower are set to have a maximum magnetic field difference of 810 G that corresponds to a maximum deceleration of the lithium atoms by 761 m/s.

The laser systems, used here, are illustrated in a modified and improved version in Sec. 3.4. The main different aspect, that is worth mentioning, are the different laser powers that influence the trapped atom numbers and MOT parameters. The MOT beam waists are therefore chosen to be smaller than the ones used later on. The laser diodes for lithium, used here (Panasonic, LNCQ05PS) with have a nominal output power of 50 mW. The diodes with higher laser power (Mitsubishi, ML101J27), that are installed later, were not available at that time. Also the laser diodes and the TA chip (M2K, model TA770) for the potassium laser system are different and

49

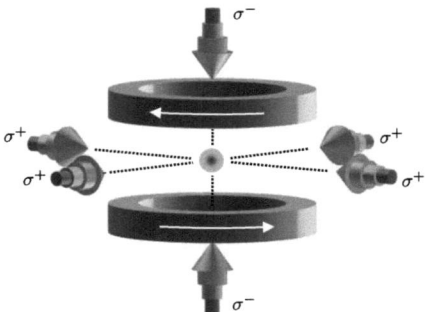

Figure 4.1: The setup of the triple MOT consists of counter-propagating beams with σ^+ and σ^- polarization, which forms a six-beam MOT. The pumping and repumping light for all three species are carefully overlapped on each other.

have a lower output power than the later used laser sources.

The atoms are detected by two methods, standard absorption imaging and fluorescence detection. In the first case, the beams for absorption imaging crosses the vertical MOT beams by a small angle and are guided onto a CCD camera (Apogee, AP1E). In the second case, a lens collects the fluorescence light of the MOTs on three photodiodes. Spectral filtering of the fluorescence light is ensured by edge filters and optical band passes (for details see (Voigt, 2004)). The filtering suppresses cross-talk between the three photodiodes below noise level. The fluorescence signal is calibrated by absorption imaging. Monitoring a MOT by fluorescence light has the advantage to observe the atom number of all three species simultaneously. Furthermore, the time loading process of a MOT can be monitored easily.

4.2 The triple MOT

A triple MOT of three species exhibits many different, independent parameters. The optimization of such a MOT is performed in view of sympathetic cooling two comparatively small fermionic clouds by a large bosonic one. This means, that for ^{87}Rb optimal parameters for a large ^{87}Rb atom number and low temperature are desired.

4.2.1 Single MOTs

In the first step, the atom number of each species is optimized in single MOT operation by varying the magnetic field gradient and the detunings of the trapping and repumping light. The $1/e^2$ beam radii of the six MOT beams are kept constant in the following and are set to 7.5 mm for lithium, 10 mm for potassium and 13 mm for rubidium. It is found that the rubidium atom number is maximum at a magnetic field gradient of $\partial B_z/\partial z = 16$ G/cm. Whereas the lithium and potassium numbers were insensitive over a wide range of magnetic field gradients. In terms of trapping the three species simultaneously, these are good news, because the atoms have to share a common field gradient. The optimal characteristic parameters of the single MOTs are summarized in Tab. 4.1.

4.2 The triple MOT

	^{6}Li	^{40}K	^{87}Rb
$\lambda_{D_2,\text{ vac}}$ (nm)	670.977	766.701	780.241
$\Gamma/2\pi$ (MHz)	5.87	6	6.07
I_{sat} (mW cm^{-2})	2.54	1.80	1.67
$I_{\text{trap}}/I_{\text{sat}}$	0.7	4	8
$I_{\text{repump}}/I_{\text{sat}}$	0.8	1.1	0.5
$\Delta\omega_{\text{trap}}$ (Γ)	-4.3	-4.2	-4.8
$\Delta\omega_{\text{repump}}$($\Gamma$)	-4.3	0	0
N_{single}	4.2×10^7	2.6×10^7	5.6×10^9
N_{triple}	3.2×10^7	1.5×10^7	5.4×10^9
T (µK)	900	40	800 (50)

Table 4.1: Characteristic parameters of three MOTs. The rubidium molasses temperature is given in brackets.

4.2.2 Triple MOT

In the second step, the parameters for triple MOT configuration are optimized by starting from the optimal parameters for single MOT configuration. Within measurement accuracy, the optimal detunings for maximum atom numbers are found to be the same as in the single MOT case. The achieved atom numbers of simultaneous trapping three species in a MOT is given in Tab. 4.1. The achieved atom numbers in the triple MOT are $N_{\text{Li}} = 3.2\times 10^7$ for ^{6}Li, $N_{\text{K}} = 1.5\times 10^7$ for ^{40}K and $N_{\text{Rb}} = 5.4\times 10^9$ for ^{87}Rb. It can be seen, that compared to the single MOT numbers, the lithium and potassium numbers dropped down by 24 % and 42 %, respectively. Whereas the rubidium atom number does not have such a significant reduction. A discussion of these losses can be found below.

4.2.3 Dispenser currents

Other crucial parameters, that are studied, are the dispenser current for potassium and rubidium. Increasing the current gives a higher partial pressure and thus a higher loading rate. On the other hand, losses due to collisions with the background gas increase. In addition, an increased dispenser current for one species always results in faster losses of the other two species. Therefore, a compromise has to be found for an optimal triple MOT operation.

To study the dispenser influence more in detail, a "clean" setup is used by loading only lithium in the single MOT. Now, the dispenser currents of rubidium and potassium are varied and the steady-state atom number of lithium is monitored. For typical rubidium dispenser currents (0 to 3.3 A) only a weak effect on lithium is observed, whereas a reduction of atom number of up to a factor of three is observed for typical potassium currents (0 to 5.2 A). Thus, as the vacuum system was extended with the UHV chamber, a second generation of potassium dispenser with a higher enrichment and an improved fabrication process were installed (see Sec. 3.3.1).

4. MAGNETO-OPTICAL TRAPPING OF THREE ATOMIC SPECIES

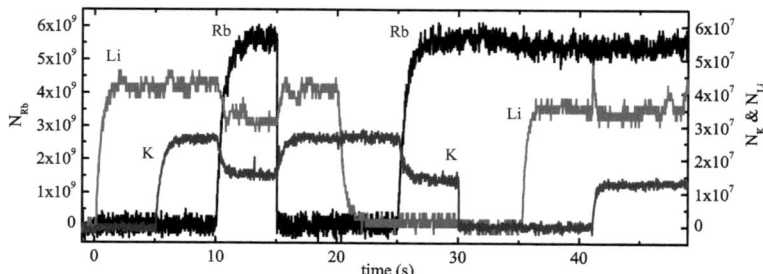

Figure 4.2: Loading sequence of the three-species MOT of ^{6}Li, ^{40}K & ^{87}Rb. The atom numbers of all possible trapping combinations between the three species are shown. Pumping and repumping light of lithium is always on and loading (blocking) of lithium is done by (un)blocking the atomic beam shutter of the Zeeman slower. For rubidium and potassium only the repumping light is (un)blocked. Between 10 s and 15 s, for example, all three species are trapped simultaneously.

Later, as the magnetic transport and finally quantum degeneracy were achieved (see Cha. 5), the dispenser currents were fine adjusted by maximizing the atom numbers in quantum degeneracy.

4.2.4 Light-assisted collisions

As discussed above, the maximum atom numbers differ in single MOT and triple MOT configuration. The extra loss in presence of other species is attributed to light-assisted collisions. For MOTs two processes are dominant: fine-structure changing collisions and radiative redistributions (Gallagher and Pritchard, 1989; Julienne and Vigué, 1991). In the following, light-assisted collisions between the species are further characterized.

For this purpose, the loading behavior of the MOT is studied in detail. Fig. 4.2 shows a MOT loading sequence of all possible trapping combinations between the three species. Here, pumping and repumping light of lithium is always on and loading (blocking) of lithium is done by (un)blocking the atomic beam shutter of the Zeeman slower. For rubidium and potassium only the repumping light is (un)blocked. These loading curves show a reduction of ^{6}Li and ^{40}K in the presence of ^{87}Rb. The curves do not show any influence of ^{6}Li on ^{40}K or vice versa. Also ^{6}Li and ^{40}K does not show any effect on ^{87}Rb.

The loading rate equation for species i in a two-species MOT with species j is given by (see also review article on cold and ultracold collisions (Weiner et al., 1999)):

$$\frac{dN_i}{dt} = L_i - \gamma_i N_i - \beta_i \int n_i^2 \, dV - \beta_{i,j} \int n_i n_j \, dV \, . \tag{4.1}$$

Here, N_i is the atom number, n_i is the atomic density, L_i is the capture rate, γ_i is the background collision rate, β_i is the intraspecies and $\beta_{i,j}$ is the interspecies loss rate. The last term of this equation describes the observed loss in atom number. In our case, the relevant interspecies loss rates are $\beta_{K,Rb}$ and $\beta_{Li,Rb}$. The intraspecies loss rates for ^{6}Li and ^{40}K can be neglected due to the comparatively small densities. The two interspecies coefficients can than be easily obtained from the loading curves: $\beta_{K,Rb} = 1 \times 10^{-11}\,\text{cm}^3\text{s}^{-1}$ and $\beta_{Li,Rb} = 8 \times 10^{-12}\,\text{cm}^3\text{s}^{-1}$, which gives the order of magnitude.

4.2.5 Optical molasses

For future experiments the temperatures of the gases are an important aspect. Optical molasses cooling, first demonstrated in (Chu et al., 1985), allows to further cool the atomic gas (see also (Dalibard and Cohen-Tannoudji, 1989; Ungar et al., 1989; Chu, 1992; Metcalf and van der Straaten, 2002)). The achieved temperatures in the triple MOT are discussed in the following.

To achieve low temperatures with optical molasses cooling, the ambient magnetic field has to be tuned to zero properly. This is done by the following scheme: Rubidium is sub-Doppler cooled by polarization gradient cooling. Varying the ambient magnetic field in one direction gives a minimum temperature for a certain magnetic field. This is followed by the other two directions. This scheme is then iterated for several times until the temperature is not further improved.

With polarization gradient cooling rubidium can be lowered in temperature from typically 800 µK in the MOT to 50 µK in only 2 ms. The MOT temperature of potassium is typically 40 µK and could not be lowered significantly by optical molasses cooling. For lithium molasses cooling does not work due to the unresolved hyperfine state (see Sec. 3.4.1). Typical lithium temperatures in the MOT are 900 µK. The short cooling time for rubidium is an essential result, because during the cooling stage the magnetic field gradient is switched-off and the clouds freely expand. For the magnetic transfer the clouds have to be recaptured and the final temperature will be increased by a larger cloud size.

4.3 Further optimization and conclusions

After extending the vacuum system, the lithium and potassium laser systems are significantly improved in optical power.

The new optimized MOT settings use a magnetic field gradient of $\partial B_z/\partial z = 15\,\text{G/cm}$. For lithium the detunings for trapping and repumping light are both $-32\,\text{MHz}$ with optical peak intensities of 0.6 and $0.5 \times I_{\text{sat}}$, respectively. For potassium the trapping light has a detuning of $-31\,\text{MHz}$ and the repumping light is detuned by $-15\,\text{MHz}$. The corresponding peak intensities are 3.3 and $0.8 \times I_{\text{sat}}$. In the rubidium case the trapping light has a detuning of $-23\,\text{MHz}$ and the repumping light is resonant. Here, the peak intensities are 8 and $0.2 \times I_{\text{sat}}$, respectively. In single MOT operation, the lithium atom number is up to $N_{\text{Li}} = 3 \times 10^9$ due to the increased optical power results. This is almost two orders of magnitude larger than the old value. In triple MOT configuration, typical atom numbers are $N_{\text{Li}} = 5 \times 10^8$ for ^{6}Li, $N_K = 1 \times 10^7$ for ^{40}K and $N_{\text{Rb}} = 3 \times 10^9$ for ^{87}Rb.

In conclusion, simultaneous magneto-optical trapping of three species is demonstrated for the first time. Losses due to light-assisted collisions are shown to be tolerable for further experiments. The achieved atom numbers and temperatures are excellent starting conditions to transfer the atoms magnetically (Greiner et al., 2001) into a UHV chamber to sympathetic cool the Fermi-Fermi mixture by the bosonic rubidium cloud into quantum degeneracy.

4. Magneto-optical trapping of three atomic species

Chapter 5
Quantum degeneracy

In this chapter I present the first quantum degenerate mixture of two fermionic species, ^{6}Li and ^{40}K. They are sympathetically cooled by a bosonic ^{87}Rb gas. In addition, a quantum degenerate two-species Fermi-Fermi mixture coexisting with a BEC is realized, representing simultaneous degeneracy of three species for the first time. Furthermore, an increased cooling efficiency for ^{6}Li by ^{87}Rb in the presence of ^{40}K is observed, demonstrating catalytic cooling.

In the following, I will give an overview on the experimental way to achieve a quantum degenerate Fermi-Fermi mixture and show crucial experimental challenges. This chapter consists of two main parts: The first one, Sec. 5.1, describes the detailed preparation of the three species for final sympathetic cooling; the second one, Sec. 5.2, is about sympathetic cooling the ensemble into quantum degeneracy.

Parts of this chapter are published in:
M. Taglieber, A.-C. Voigt, T. Aoki, T. W. Hänsch, and K. Dieckmann, "Quantum Degenerate Two-Species Fermi-Fermi Mixture Coexisting with a Bose-Einstein Condensate," Phys. Rev. Lett. **100**, 010401 (2008).

5.1 On the road to quantum degeneracy

Starting from the triple MOT, this section describes the preparation of cold and dense clouds for sympathetic cooling in the QUIC trap. Optimal starting conditions for the individual parameters of the different species, often contradicting each other, are found and discussed.

Compared to the previous chapter, the experimental setup is now extended by the UHV chamber, the magnetic transfer and the QUIC setup (see Cha. 3 for details).

5.1.1 cMOT and dMOT

For starting sympathetic cooling, clouds with high phase space densities are desired. Thus, after simultaneous trapping the species in a MOT, the atoms are cooled and compressed further, which is also relevant for optimal loading the atoms into the quadrupole trap with subsequent magnetic transfer. Here, I present a suitable strategy that considers the different atomic properties.

As seen before in Cha. 4, lithium can not be sub-Doppler cooled with standard techniques and the achieved temperature in the MOT of typically 900 µK is comparatively high. In combination with its small mass lithium will expand much faster during free expansion, necessary

5. QUANTUM DEGENERACY

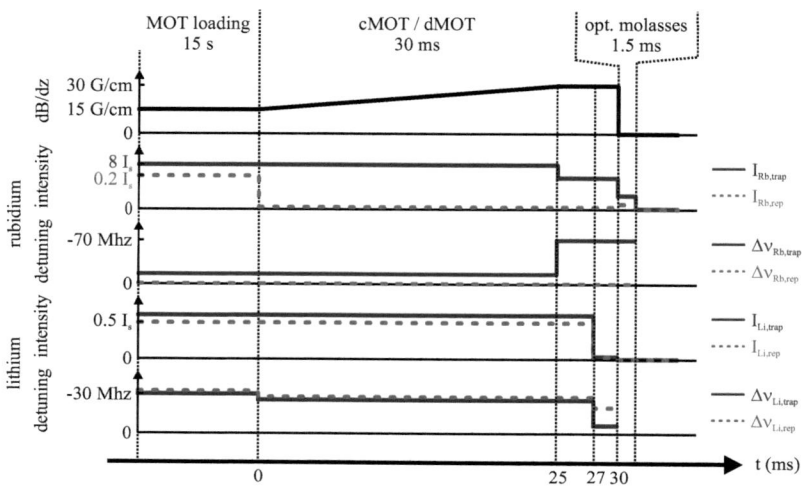

Figure 5.1: Experimental sequence of the combined cMOT and dMOT for lithium and rubidium, respectively. First, the triple MOT is loaded with maximum atom numbers within 15 s. Then the magnetic field gradient of the MOT is increased and the ^{6}Li and ^{87}Rb clouds are cooled and compressed by a cMOT and dMOT phase within 30 ms. After the cMOT/dMOT scheme the magnetic field gradient is switched off and rubidium is further cooled by polarization gradient cooling within 1.5 ms.

for sub-Doppler cooling of rubidium. This makes it difficult to find an appropriate magnetic capture gradient for loading the clouds into the quadrupole trap without loosing significantly in phase space densities of all three species. Due to the small interspecies scattering length between ^{6}Li and ^{87}Rb ($\sigma_{\text{Li,Rb}} = -19.8\,a_0$) (Li et al., 2008), the initial phase space density of lithium is a crucial parameter and has to be increased for efficient sympathetic cooling.

To overcome these difficulties, a combination of a compressed MOT (cMOT) (Petrich et al., 1994) for ^{6}Li and a temporal dark MOT (dMOT) (Adams et al., 1995; Friebel et al., 1998) and molasses phase for ^{87}Rb are implemented in the experiment. The basic idea here is to first load the triple MOT with maximum atom numbers. Then the MOT gradient is ramped up and the ^{6}Li and ^{87}Rb clouds are cooled and compressed by a cMOT and dMOT phase, respectively. After switching off the field gradient, only ^{87}Rb is further cooled through a molasses phase, which is now shorter due to the dMOT phase. This allows to load the clouds into the quadrupole trap with higher phase space densities. In addition, the transfer efficiency of ^{6}Li through the narrow differential pumping tube is increased.

The details of the cMOT/dMOT scheme is illustrated in Fig. 5.1. In the following, the presented densities and temperatures correspond to single-species MOT operations. The MOTs are loaded with a magnetic field gradient of $\partial B_z/\partial z = 15\,\text{G/cm}$ within 15 s. The magnetic field of the lithium Zeeman slower is switched off 50 ms just before the MOT loading is finished. This is enough time for the clouds to readjust to the new magnetic field configuration and to prevent stray magnetic fields during the subsequent optical molasses phase. The gradient is then increased to $\partial B_z/\partial z = 30\,\mathrm{G/cm}$ within 25 ms. During this ramp the rubidium repumper light is reduced to 1 % saturation intensity. After that the detuning and intensity of the rubidium pumping light are reduced to $-72\,\mathrm{MHz}$ and 70 % of the initial value, respectively. The rubidium cloud is then cooled and compressed within 5 ms to typically $200\,\mathrm{\mu K}$ and $3\times10^{11}\,\mathrm{cm^{-3}}$ compared to $750\,\mathrm{\mu K}$ and $0.4\times10^{11}\,\mathrm{cm^{-3}}$ before the dMOT. On the other hand, as the magnetic gradient field starts to increase, the lithium cMOT begins with detunings of $-28\,\mathrm{MHz}$ for the repumping and pumping light. Note, that these values might be changed during the normal MOT phase to reduce the lithium heat load on rubidium. At the end of the sequence the lithium detunings are shifted closer to resonance for 3 ms, $-11\,\mathrm{MHz}$ and $-21\,\mathrm{MHz}$ for pumping and repumping light. Both intensities are decreased to 6 % saturation intensity. The lithium cMOT increase the density and lowers the temperature from typically $1.3\,\mathrm{mK}$ and $0.9\times10^{10}\,\mathrm{cm^{-3}}$ to $500\,\mathrm{\mu K}$ and $1.5\times10^{10}\,\mathrm{cm^{-3}}$.

After the cMOT/dMOT scheme the magnetic field gradient is switched off and rubidium is further cooled to typically $45\,\mathrm{\mu K}$ in only 1.5 ms. For this the rubidium laser detunings are kept the same as during the dMOT phase and the intensity of the pumping light is further decreased to 30 % of the initial MOT intensity. The repumping light is only slightly increased to 2 % saturation intensity. In total, the combined cMOT/dMOT schemes increases the phase space density of rubidium by almost to orders of magnitude and of lithium by a factor of 5.

The scheme and values given above are chosen such that the reliability is high and the necessary number of parameters are small. Hence, a molasses phase is neither implemented for potassium nor for lithium. The lithium cMOT requires exact alignment and intensity balancing of the six MOT beams, which become more sensitive parameters as the laser detunings are ramped close to resonance. Thus, to guarantee a reliable operation over several days, the cMOT laser detunings are chosen to be slightly more red-detuned than the optimal detunings for minimal temperature.

5.1.2 State preparation

To trap the three species $^{87}\mathrm{Rb}$, $^{40}\mathrm{K}$ and $^{6}\mathrm{Li}$ magnetically, the atoms have to be prepared into low-field seeking states that are stable in the three species mixture. Usually, the fastest inelastic collision process for alkali atoms is due to spin-exchange collisions that result in depolarization and losses. For this case the selection rules are given by (Tiesinga et al., 1993)

$$\triangle l = 0 \qquad (5.1)$$
$$\triangle m_l = 0 \qquad (5.2)$$
$$\triangle M_\mathrm{F} = 0 \,. \qquad (5.3)$$

Here, l is the relative angular momentum of two atoms, m_l is the corresponding projection quantum number and M_F is the projection quantum number of the total spin $\mathbf{F} = \mathbf{f}_1 + \mathbf{f}_2$ of two atoms at large interatomic distance with the hyperfine quantum numbers $|f_1, m_{f1}\rangle$ and $|f_2, m_{f2}\rangle$. In the three species mixture there exists only one magnetic trappable combinations of states, which is stable against spin-exchange collisions:

$$^{87}\mathrm{Rb}\,|2,2\rangle \,\&\, ^{40}\mathrm{K}\,|9/2,9/2\rangle \,\&\, ^{6}\mathrm{Li}\,|3/2,3/2\rangle \qquad (5.4)$$

This state can still decay through dipolar relaxation (Mies et al., 1996; Moerdijk and Verhaar, 1996; Boesten et al., 1996), which is typically on the order of $10^{-15}\,\mathrm{cm^3 s^{-1}}$ for alkali atoms

5. QUANTUM DEGENERACY

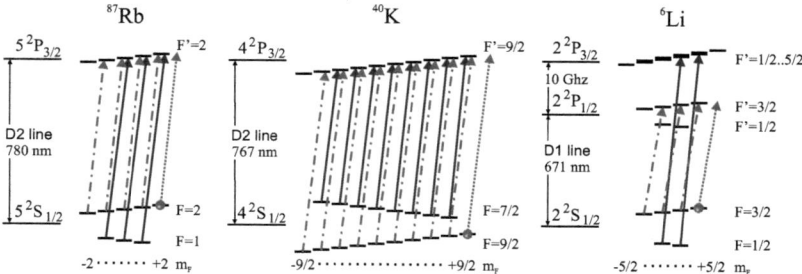

Figure 5.2: State preparation of ^{87}Rb, ^{40}K and ^{6}Li. The atoms are prepared into dark states by optical pumping with σ^+-light (dashed lines). Additional repumper light (solid lines) brings the atoms back into the pumping transition. ^{87}Rb is pumped into the $|F = 2, m_F = 2\rangle$ state, ^{40}K into the $|9/2, 9/2\rangle$ state and ^{6}Li into the $|3/2, 3/2\rangle$ state. ^{87}Rb and ^{40}K use the D_2-line, whereas ^{6}Li is optically pumped on the D_1-line.

and becomes only important in the high density regime. For a detailed discussion I refer to (Taglieber, 2008).

The mixture (5.4) is prepared by optical pumping. For this purpose, σ^+-light pumps the atoms into the appropriate dark states, which is shown for ^{87}Rb, ^{40}K and ^{6}Li in Fig. 5.2. Optical pumping for ^{87}Rb and ^{40}K is performed on the D_2-lines, $|F = 2\rangle \rightarrow |F' = 2\rangle$ for ^{87}Rb and $|F = 9/2\rangle \rightarrow |F' = 9/2\rangle$ for ^{40}K. In the ^{6}Li case optical pumping into a dark state is not efficient because of the unresolved hyperfine structure of the first exited state. Therefore, the transition $|F = 3/2\rangle \rightarrow |F' = 3/2\rangle$ on the D_1-line is used, which requires an additional laser to the MOT and Zeeman lasers. During optical pumping repumper light brings the atoms back into the pumping transition.

The time sequence for optical pumping is as follows: After the molasses phase for ^{87}Rb a bias field is switched on to about 9 G. This relative high field reduces imperfect polarization effects of the optical pumping beam. ^{87}Rb and ^{6}Li are then optically pumped within 160 µs for ^{87}Rb and 300 µs for ^{6}Li into the desired states. The total intensities are typically 2 mW/cm^2 for ^{87}Rb and 1 mW/cm^2 for ^{6}Li. After that the bias field is reduced to about 1 G to pump ^{40}K into ^{40}K $|9/2, 9/2\rangle$. The lowered magnetic field ensures to address all states optically, see (Eigenwillig, 2007) for details. The pulse duration for ^{40}K is 100 µs and the total light intensity is 2 mW/cm^2.

5.1.3 Magnetic transport

The magnetic transfer of the atoms into the UHV chamber starts with recapturing the atoms magnetically after optical pumping. The magnetic capture gradient for maximal phase space density is a function of the initial temperature and density distribution. For different atomic species the optimal gradient is in general not the same and a compromise has to be made. For the following experiments, described within this thesis, a capture gradient of $\partial B_z/\partial z = 140$ G/cm is chosen and is kept constant during the magnetic transfer. Due to the comparatively small mass and high temperature of lithium, the lithium cloud is about ten times larger than the rubidium and potassium clouds. Thus, larger gradients allow efficient transfer of the lithium cloud through the narrow differential pumping tube.

The transfer sequence of the quadrupole coils is experimentally optimized. The $1/e$ lifetimes of the clouds are 500 ms in the MOT chamber and 60 s in the class cell in the UHV chamber for typical dispenser currents. Thus, fast extraction of the clouds out of the MOT chamber is necessary. The final implemented sequence for the first part of the transfer (from the MOT position to the corner) is a compromise between heating and atom loss due to background gas collisions on the one hand side and heating due to non-adiabatic acceleration of the clouds on the other hand side. The total time, used to transfer the atoms to the corner, is 1.0 s. The second part of the transfer (from the corner to the glass cell) is much less critical. Here, the optimal transfer time is 1.5 s. For larger times no further improvement in the final temperature is observed.

5.1.4 QUIC trap

Evaporative cooling is performed in the QUIC trap, which suppresses Majorana losses. This section briefly describes the loading sequence and characterization of the QUIC trap.

Loading sequence The loading sequence is based on three steps:

1. The clouds are adiabatically compressed in the quadrupole trap. The compression goes from $\partial B_z/\partial z = 140\,\text{G/cm}$ to $300\,\text{G/cm}$ in 2.0 s.

2. The atoms are transferred into the QUIC trap in 3.0 s.

3. The magnetic bias field is increased from 1.2 G to 3.2 G in 300 ms. This further suppresses Majorana losses, which are dominant for ^{6}Li.

QUIC characterization The QUIC trap can be fully characterized by three parameters: the bias field B_0, the radial and axial trapping frequencies.

The bias field or trap bottom is measured by Zeeman or hyperfine transitions on ^{87}Rb. A rubidium cloud of a few µK is prepared by evaporative cooling in the QUIC trap. The RF power is reduced to decrease dressing of the states. Then atom loss is monitored as a function of RF frequency. With this technique the trap bottom is determined to $B_0 = 3.16(3)$ G.

The trapping frequencies are determined by center-of-mass dipole oscillations. This is done by displacing a ^{40}K cloud by adding a sinusoidal magnetic field along the radial or axial directions (for details on the setup and parameters see (Eigenwillig, 2007)). The measured frequencies for ^{40}K are $\omega_x^K = 2\pi \times 29.97(3)$ Hz and $\omega_\rho^K = 2\pi \times 230.88(5)$ Hz along the axial and radial directions, respectively. With the trap bottom and trapping frequencies the α and β values of the QUIC trap (see Sec. 3.6.3) are determined to $\alpha = 146.1(7)$ G/cm and $\beta = 254.0(5)$ G/cm^2.

The other trapping frequencies for ^{6}Li and ^{87}Rb are easily calculated through the relations $\omega^{Li}/\omega^K = \sqrt{m_K/m_{Li}} \approx 2.58$ and $\omega^{Rb}/\omega^K = \sqrt{m_K/m_{Rb}} \approx 0.678$. This gives $\omega_x^{Li} = 2\pi \times 77.39(8)$ Hz and $\omega_\rho^{Li} = 2\pi \times 596.1(1)$ Hz for ^{6}Li and $\omega_x^{Rb} = 2\pi \times 20.32(2)$ Hz and $\omega_\rho^{Rb} = 2\pi \times 156.55(3)$ Hz for ^{87}Rb.

5. Quantum degeneracy

5.2 Cooling into quantum degeneracy

In the QUIC trap, the two fermionic species ^{6}Li and ^{40}K are sympathetically cooled by the bosonic ^{87}Rb gas into quantum degeneracy. Below, I describe the experimental method and challenges to reach this regime.

5.2.1 Evaporative cooling

Cooling a bosonic gas into a BEC is achieved by evaporative cooling, which so far is the only experimental way. Current laser cooling schemes are limited by reabsorption processes (Walker et al., 1990) and inelastic, light-induced collisions (Walker and Feng, 1994; Weiner, 1995). Evaporative cooling was first demonstrated for hydrogen (Hess, 1986; Masuhara et al., 1988) and the first BECs using RF induced evaporation (Pritchard et al., 1989) were realized with alkali atoms in 1995 (Anderson et al., 1995; Davis et al., 1995; Bradley et al., 1995). Extensive descriptions on evaporative cooling can be found in (Ketterle and van Druten, 1996; Luiten et al., 1996; Walraven, 1996).

5.2.1.1 Principle

The basic principle of evaporative cooling is the following: Take a gas in thermal equilibrium with temperature T_i; remove high energetic particles above a certain truncation energy E_{cut}; due to elastic collisions atoms are continuously redistributed to higher energies and hence lost from the trap, while the gas relaxes to a lower temperature $T_f < T_i$ ("plain evaporation"); constantly reducing E_{cut} maintains the evaporation process (forced evaporation).

A parameter that characterizes the efficiency of evaporation is given by (Ketterle and van Druten, 1996)

$$\chi = -\frac{d(\ln D)}{d(\ln N)} = -\frac{\dot{D}/D}{\dot{N}/N}. \tag{5.5}$$

Here $D = n\lambda_{\text{dB}}^3$ is the phase space density. Maximizing χ at any point optimize the whole evaporation process (Ketterle and van Druten, 1996).

In typical experiments the phase space density is increased by more than six orders of magnitude, making evaporative cooling a powerful tool.

A successful evaporation process requires that non-evaporative loss rates due to inelastic collisions ("bad losses") are smaller than a certain value. The relevant expression is here the "ratio of good to bad collisions" $1/\lambda$. For certain ratios the evaporation process can be performed in the "run-away" regime, i.e. that in spite of cooling the density in the trap increases so much that the elastic collision rate increases and therefore the evaporation is speeded up. It can be shown that this regime is accessible for typical traps with $1/\lambda$ larger than $\approx 10^2$ (Valkering, 1999).

5.2.1.2 Evaporative cooling in mixtures

In magnetic traps, the standard evaporation scheme uses RF induced transitions due to its simplicity. Here, for sufficiently high RF amplitude, the RF field drives transitions from magnetically trapped to untrapped states within one hyperfine manifold. The resonance condition in the linear Zeeman regime is given by $\hbar\omega_{\text{rf}} = \mu_B g_F |B(\mathbf{r})|$. The corresponding truncation energy for atomic states m_F is then $E_{\text{cut}} = m_F \hbar\omega_{\text{rf}} - \mu_B B_0$, where B_0 is the magnetic field

5.2 Cooling into quantum degeneracy

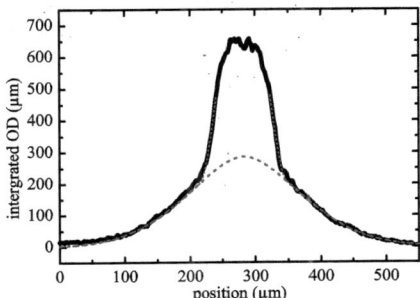

Figure 5.3: A typical bimodal density distribution of the thermal and condensed part of a rubidium cloud. The graph shows the integrated optical density distribution of ^{87}Rb after a time-of-flight of 20 ms. A two-component fit gives an atom number of $N_{\text{thermal}} = 2.2 \times 10^5$ for the thermal component and $N_0 = 1.3 \times 10^6$ for the condensate. The temperature is fitted to 214 nK.

at the trap center. Note, that due to the quadratic Zeeman effect, which is relevant at high magnetic fields, the evaporation process could be incomplete (Desruelle et al., 1999) and could lead to the formation of an Oort cloud (Dieckmann, 2001).

For mixtures, RF evaporation can also lead to unwanted transitions and losses of the other atoms. This is the case in the three-species mixture of ^{87}Rb $|2, 2\rangle$, ^{40}K $|9/2, 9/2\rangle$ and ^{6}Li $|3/2, 3/2\rangle$, where due to the different Landé factors the respective truncation energy scales differently with the magnetic field. For clouds of the same temperature it can easily be seen that evaporation of ^{87}Rb also evaporates ^{40}K at a higher truncation energy and ^{6}Li at a lower one. For experiments with a ^{87}Rb - ^{40}K mixture, ^{40}K is sympathetically cooled and in thermal equilibrium with ^{87}Rb. Thus, ^{40}K is not affected by RF evaporation and quantum degeneracy can be achieved, as shown in other experiments (Roati et al., 2002; Goldwin et al. 2004; Ospelkaus et al., 2006b; Rom et al., 2006). In a ^{87}Rb-^{6}Li mixture on the other hand, first ^{6}Li will be removed and then ^{87}Rb by lowering the RF cut. This contradicts the concept of sympathetic cooling of ^{6}Li by ^{87}Rb that is based on sympathetic cooling the two fermionic clouds by ^{87}Rb with minimal atom loss of the fermions.

In order to avoid the above mentioned undesired removal of atoms from the trap ^{87}Rb is evaporated by MW induced transitions (see Sec. 5.2.1.4). This method is state selective and drives only transitions of ^{87}Rb due to the large differences of the hyperfine splittings, which are 6.8 GHz for ^{87}Rb, 1.3 GHz for ^{40}K and 228 MHz for ^{6}Li (see also Appendix A.2).

5.2.1.3 First BEC by RF evaporation

To demonstrate the functionality of our vacuum chamber, the first BEC in our experiment is realized by RF evaporation due to its relative simple implementation. As mentioned above, RF evaporation can also be used for the ^{87}Rb-^{40}K mixture. We also successfully realized double quantum degeneracy of this Bose-Fermi combination by applying this scheme.

5. QUANTUM DEGENERACY

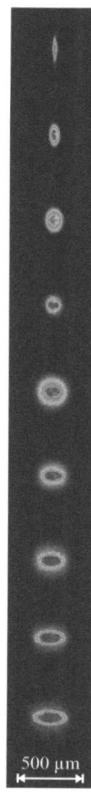

Figure 5.4: Anisotropic expansion of a Bose-Einstein condensate. The time-of-flight is varied top down from 0 to 32 ms in steps of 4 ms.

To optimize the RF evaporation sequence the vacuum is set to typical vacuum conditions, i.e. all atomic sources (lithium oven and dispensers) are switched on to typical values. The measured loss rate in the QUIC trap is $\tau_{\text{loss}} = 60\,\text{s}$. The starting conditions in the QUIC with 6×10^8 rubidium atoms and a temperature of 620 µK gives a elastic collision rate of $\tau_{\text{el}}^{-1} = 19\,\text{s}^{-1}$. The ratio of good to bad collisions is then larger than 10^3 and run-away evaporation is possible. The optimal evaporation sequence, which is used to produce a BEC, consists of nine adjacent linear frequency ramps and each ramp is optimized in evaporation efficiency. The whole sequence has an efficiency of $\chi = 3.03\,(7)$, i.e. the phase-space density increase by about three orders of magnitude for one order of magnitude particle number decrease. The sequence starts at 50 MHz and stops at 2.2 MHz, where BEC is reached. To avoid dressing of the states, which gives smaller trapping frequencies, the RF power is gradually decreased to the end. At the end of the evaporation sequence the RF frequency is slightly increased and the gas is allowed to thermalize within 40 ms.

The evidence of BEC is shown in Fig. 5.3. It shows the bimodal density distribution of ^{87}Rb after a time-of-flight of 20 ms. A two-component least square fit gives an atom number of $N_{\text{thermal}} = 2.2 \times 10^5$ for the thermal component and $N_0 = 1.3 \times 10^6$ for the condensate. The temperature is fitted to 214 nK.

Another characteristic signature of BEC is the anisotropic expansion, shown in Fig. 5.4. To demonstrate this the time-of-flight is varied from 0 to 32 ms in steps of 4 ms and clearly shows the inversion of ellipticity, as given by Eq. (2.21) and (2.22).

5.2.1.4 MW evaporation

Compared to RF induced transitions between Zeeman states, MW evaporation drives hyperfine transitions and is therefore a species selective method. In the MW case, the rubidium evaporation is carried out on the $|2, 2\rangle \rightarrow |1, 1\rangle$ transition, which is at $\nu_{\text{HF,Rb}} \approx 6.8\,\text{GHz}$. The corresponding frequencies for the MW evaporation sequence are calculated from the previous optimized RF evaporation sequence. It starts at $119\,\text{MHz} + \nu_{\text{HF,Rb}}$ and ends at $6.7\,\text{MHz} + \nu_{\text{HF,Rb}}$ above the trap bottom.

5.2 Cooling into quantum degeneracy

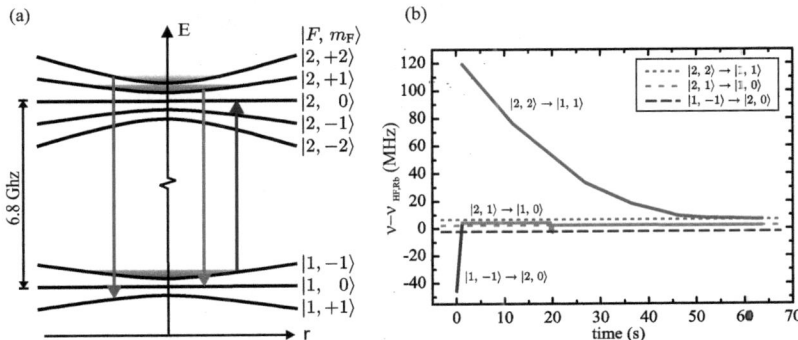

Figure 5.5: (a) Hyperfine transitions for species selective evaporative cooling of rubidium. The evaporation is performed on the $|2,2\rangle \to |1,1\rangle$ transition. Additional cleaning sequences are done on the $|2,1\rangle \to |1,0\rangle$ and $|1,-1\rangle \to |2,0\rangle$ transitions to remove atoms in unwanted states. (b) Evaporation and cleaning sequence. The dashed lines represent the resonant transitions at the trap center.

Experimentally it is found that the $|2,1\rangle$ state continuously repopulates during evaporation and that $|2,1\rangle$ state cleaning is an important precondition for BEC and degenerate fermions. Thus, in addition to the evaporation frequency, the $|2,1\rangle$ and also the $|1,-1\rangle$ state are cleaned during evaporation to avoid depolarization and heating due to spin-exchange collisions. Moreover, occupied $|1,-1\rangle$ and $|2,1\rangle$ states would represent additional heat load and, therefore, have to be removed. The relevant frequencies are illustrated in Fig. 5.5 (a). The main process of the repopulation of the $|2,1\rangle$ state is attributed to the evaporation that drives $|2,2\rangle \to |1,1\rangle$ transitions. When atoms in the $|1,1\rangle$ state are formed, they are not magnetically trapped and accelerated outwards of the trap center. At a certain point these atoms become resonant again with the evaporation frequency, but now the MW drives the transition $|1,1\rangle \to |2,1\rangle$. To prevent such a π-transition different antenna geometries are investigated, for example a helix antenna, but no suppression of the undesired transition is found. This could be due to the not well-defined quantization axis for high temperatures in the magnetic trap. In addition, the $|2,1\rangle$ state is repopulated on a slower timescale by dipolar relaxation of the $|2,2\rangle$ state. Due to the continuous repopulation of the $|2,1\rangle$ state, a second MW frequency is implemented that continuously removes the $|2,1\rangle$ state (see also MW design in Sec. 3.9). The sequence of this cleaning is optimized with a Stern-Gerlach method that allows state selective detection of the different rubidium states (see Sec. 3.5).

The sequence for evaporation of ^{87}Rb by hyperfine transitions is based on the previous optimized RF sequence. To sympathetic cool ^{6}Li and ^{40}K simultaneously, the sequence is stretched in time (see discussion in Sec. 5.2.2.3). The detailed evaporation and cleaning sequence of ^{87}Rb in the stretched version is shown in Fig. 5.5 (b). First, $|1,-1\rangle$ states are removed by a $|1,-1\rangle \to |2,0\rangle$ transition within 1.2 s. The sequence starts 43 MHz below the $|1,-1\rangle \to |2,0\rangle$ transition at the trap center and stops 2 MHz above it. This sweep, which has discrete steps due to technical limitations, is done by the MW synthesizer (Systron Donner) only and the two

5. QUANTUM DEGENERACY

sidebands are switched off (see Sec. 3.9). Then, the first sideband for evaporation is switched on and the evaporation sequence starts. Cleaning of the $|2, 1\rangle$ state is done by the MW carrier frequency. Following, at 20 s, the carrier frequency is lowered by -50 MHz. The second sideband for cleaning is switched on and the $|1, -1\rangle$ atoms are removed by a continuous frequency sweep with a span of 1 MHz. Finally, the $|2, 1\rangle$ state is cleaned by a linear ramp with a span of 150 kHz.

5.2.2 Sympathetic cooling

The successful realization of a rubidium BEC through MW evaporation marks an important precondition for sympathetic cooling of potassium and lithium into quantum degeneracy. In the following, the principle of sympathetic cooling is explained and the route to quantum degeneracy and its challenges are discussed. Finally, catalytic cooling and a quantum degenerate mixture of two fermionic species coexisting with a BEC are presented.

5.2.2.1 Principle

The principle of sympathetic cooling, which was proposed by (Wineland et al., 1978), is to cool particles that stay in thermal contact with an actively cooled bath. The first experimental demonstration was in an ion trap with charged particles that are interacting through long-range Coulomb collisions (Drullinger et al., 1980; Larson et al., 1986). Neutral atoms were first sympathetically cooled in a spin mixture of ^{87}Rb, producing two overlapping BECs (Myatt et al., 1997). Moreover, atoms and even molecules can be cooled by cryogenically cooled helium (Kim et al., 1997; Weinstein et al., 1998). The application to fermions was first demonstrated in the Fermi-Bose mixture ^{6}Li – ^{7}Li (Schreck et al., 2001), in which the bosonic cloud is evaporatively cooled. Within this thesis, the sympathetic cooling technique is applied to cool two different species (^{6}Li and ^{40}K) by a bosonic cloud (^{87}Rb) through elastic collisions.

Sympathetic cooling process In the non-degenerate classical regime, rate equations of the thermalization process between two mixtures of particles 1 and 2 can be analytically deduced. The average energy transfer per collision has the following form (Mosk et al., 2001):

$$\Delta E_{\text{coll}} = k_B \, \Delta T \, \xi \, , \text{ with } \xi = \frac{4 \, m_1 \, m_2}{(m_1 + m_2)^2} \, . \tag{5.6}$$

In this equation, $\Delta T = T_2 - T_1$ is the temperature difference between the two mixtures and ξ is a mass-dependent parameter that reduces the energy transfer. The corresponding energy exchange rate is given by $\Gamma_{12} \, \Delta E_{\text{coll}}$, where the average collision rate between the two mixtures is (Mosk et al., 2001)

$$\Gamma_{12} = \sigma_{12} \, \langle v_{12} \rangle \int n_1(\mathbf{r}) \, n_2(\mathbf{r}) \, d\mathbf{r} \, . \tag{5.7}$$

Here, $\sigma_{12} = 4\pi \, a_{12}^2$ is the interspecies cross section between distinguishable particles with the s-wave scattering length a_{12} and $\langle v_{12} \rangle$ is the mean thermal relative velocity, which can be deduced from the classical Maxwell-Boltzmann velocity distribution (Mosk et al., 2001):

$$\langle v_{12} \rangle = \left[\frac{8 \, k_B}{\pi} \left(\frac{T_1}{m_1} + \frac{T_2}{m_2} \right) \right]^{1/2} \, . \tag{5.8}$$

5.2 Cooling into quantum degeneracy

In an isolated system and assuming thermal equilibrium within each mixture, rate equations of the thermalization process can be calculated by the energy exchange rate. The thermalization rate in a two-species mixture with heat capacities C_i ($i = 1, 2$) is given by

$$\tau_{\text{therm}}^{-1} = -\frac{1}{\Delta T}\frac{d(\Delta T)}{dt} = \Gamma_{12}\,\xi\left(\frac{k_\text{B}}{C_1} + \frac{k_\text{B}}{C_2}\right). \tag{5.9}$$

In a harmonic confinement the heat capacity per atom is $3\,k_\text{B}$. For $N_1 \ll N_2$ this yields $3/\xi$ collisions per particle 1 that are needed for thermalization (Delannoy et al., 2001; Mosk et al., 2001).

Practical aspects of sympathetic cooling In practice, several aspects and limitations have to be considered:

First of all, the density overlap of the two clouds is an essential aspect, which can be seen from Eq. (5.7). At ultracold temperatures a spin-polarized Fermi gas does not thermalize due to the Pauli exclusion principle that strongly suppresses collisions (see Sec. 2.2.1.2). Therefore, each single fermionic atom has to be cooled by the cooling bath. If the initial overlap between the clouds is only local, this could lead to an incomplete cooling of the fermions: a cold, thermalized part that is surrounded by a hot halo cloud. Furthermore, the density overlap for mixtures with different trapping frequencies is reduced by the gravitational sag, which is given by $\Delta z_{\text{sag}} = -g/\omega_z^2$. Here, ω_z is the angular trapping frequency along $\mathbf{g}$. Moreover, the strength of the interspecies scattering length and its sign, whether attractive or repulsive, can influence the density overlap significantly and can even lead to a phase separation of the two clouds (Ospelkaus et al., 2006c).

Another aspect is the scattering cross section that is in general energy dependent. In ^{40}K, for example, a p-wave shape resonance at a collision energy of $280\,\mu\text{K} \times k_\text{B}$ (DeMarco et al., 1999a) enhances the collision rate. In ^{40}K – ^{87}Rb, on the other hand, the large negative s-wave scattering length leads to a vanishing s-wave scattering cross section at $630\,\mu\text{K} \times k_\text{B}$ (Goldwin, 2005) due to the Ramsauer-Townsend effect.

Furthermore, quantum statistical effects play a crucial role in the cooling process. The efficiency of sympathetic cooling is affected by the different heat capacities of bosons and fermions (Presilla and Onofrio, 2003) and by superfluidity of the BEC (Timmermans and Côté, 1998). In addition, the cloud sizes of quantum degenerate bosons and fermions differ significantly and the density overlap is reduced. Moreover, Pauli blocking (Holland et al., 2000; DeMarco et al., 2001) decelerates the cooling process for fermions at quantum degeneracy due to a reduced number of unoccupied final scattering states.

5.2.2.2 Sympathetic cooling of ^{40}K

After the successful realization of a BEC by hyperfine transitions, the short MW evaporation sequence, which is based on the RF evaporation sequence, is tried out to cool ^{40}K sympathetically. This sequence works remarkable well and a quantum degenerate ^{87}Rb-^{40}K mixture with high atom numbers and low temperature is comparatively easily achieved. This is attributed to two parameters that are relevant for the interspecies thermalization rate (see Eq. (5.9) for details). First, the relatively small interspecies mass ratio of 2.2 between ^{87}Rb and ^{40}K yields a large energy transfer parameter of $\xi = 0.86$ (see Eq. (5.6)). Second, the high s-wave scattering length of $a_{\text{K,Rb}} = -215\,(10)\,a_0$ between ^{87}Rb and ^{40}K results in a large collision rate. The negative sign even improves the density-density overlap.

65

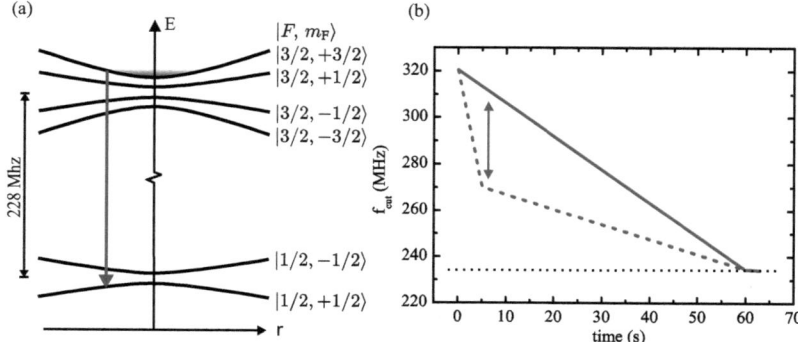

Figure 5.6: (a) The lithium hyperfine transition from $|3/2, 3/2\rangle \rightarrow |1/2, 1/2\rangle$ is used to selectively remove high energetic lithium atoms. (b) Cleaning sequence of the $|3/2, 3/2\rangle$ state (solid line). The total heat load of lithium is fine-tuned by varying the initial frequency part (dashed line). The horizontal dotted line represents the resonant transition at the trap center.

Typically, a quantum degenerate fermionic cloud of $T_K = 0.2\, T_F^K$ with $N_K = 1.4 \times 10^5$ and a BEC of $N_{0,\text{Rb}} = 3.5 \times 10^5$ are simultaneously achieved.

5.2.2.3 Sympathetic cooling of ^{6}Li

In the case of sympathetic cooling of ^{6}Li, we found a reduced thermalization rate and different challenges have to be solved to achieve quantum degeneracy of ^{6}Li. The above presented MW evaporation sequence in the simple form, which works well for cooling ^{40}K, does not work for ^{6}Li and no ^{6}Li atoms are detected after the whole evaporation sequence. Optimizing the evaporation sequence for ^{6}Li is hindered by the small ^{6}Li mass and the relatively high temperature at the beginning of the cooling process. This makes it difficult to determine temperature and atom number of the ^{6}Li cloud due to its fast expansion in a time-of-flight measurement. Typical starting conditions in the QUIC trap are $N_{\text{Li}} = 1-2 \times 10^7$ and $T_{\text{Li}} = 1.5\,(5)$ mK for ^{6}Li and $N_{\text{Rb}} = 5-8 \times 10^8$ and $T_{\text{Rb}} = 0.6\,(1)$ mK for ^{87}Rb.

The main parameters that reduce the thermalization time and complicate the sympathetic cooling are the following:

The ^{87}Rb-^{6}Li s-wave scattering length, recently determined by Feshbach spectroscopy, is $-19.8\, a_0$ (Li et al., 2008), which is more than one order of magnitude smaller than the ^{87}Rb-^{40}K scattering length. This gives a more than 100 times smaller interspecies cross section for ^{87}Rb-^{6}Li in comparison to ^{87}Rb-^{40}K. In addition, the relatively high ^{87}Rb-^{6}Li mass ratio of 14.4 ($\xi = 0.24$) leads to a lower energy transfer per collision. These two factors lead to a significant reduction of the thermalization time. The higher mean thermal relative velocity, due to the smaller mass of lithium, only partially compensates this reduction. Furthermore, the relative gravitational sag between ^{87}Rb and ^{6}Li is $\Delta z_{\text{sag}}^{\text{Li,Rb}} = 9.4\,\mu\text{m}$ in the QUIC trap and is almost a factor of two larger than between ^{87}Rb and ^{40}K ($\Delta z_{\text{sag}}^{\text{K,Rb}} = 5.5\,\mu\text{m}$). The relative sag becomes relevant for temperatures below 1 µK and for atom numbers, where the sag is comparable to

5.2 Cooling into quantum degeneracy

the cloud size. This sag reduces the density-density overlap and suppresses sympathetic cooling towards lower temperatures.

In order to achieve a degenerate fermionic cloud of lithium with high atom numbers, three methods are used that go well beyond the usual effort to obtain quantum degeneracy:

First, the evaporation sequence is clearly lengthened in time. It is found that prolonging the sequence at the end is more favorable than at the beginning. The final optimized sequence, that is used within this thesis, has a total length of 63 s and is a compromise between optimal thermalization time of lithium and losses due to background gas collisions. The evaporation sequence is shown in Fig. 5.5.

Second, after extending the evaporation sequence a cold atomic cloud of ^{6}Li is observed at the end of evaporation. This cold cloud is surrounded by a hot halo cloud of ^{6}Li atoms. Moreover, no BEC is observed. This implies that lithium is not in full thermal contact with rubidium and hot lithium atoms, which are not affected by the sympathetic cooling process, heat up rubidium by collisions. It is found, that a BEC and a significant lithium atom number with cold temperatures can be achieved by selectively removing the hot lithium part during the cooling process. This is done by driving the lithium hyperfine transition from $|3/2, 3/2\rangle \rightarrow |1/2, 1/2\rangle$ (see Fig. 5.6 (a)), which is, as in the MW evaporation case, a species selective method. The frequency of the lithium hyperfine transition is adjusted during the cooling sequence and is varied in two adjacent linear ramps (see solid line in Fig. 5.6 (b)). The first ramp, that has a duration of 60 s, starts at 320 MHz, which corresponds to a cut energy of $\sim k_B \times 3$ mK, and stops at 234.6 MHz ($\sim k_B \times 15\,\mu$K). Then, the frequency is ramped to 234.3 MHz ($\sim k_B \times 4\,\mu$K) within 3 s. Lowering the initial frequency and thus the cut energy at the beginning gives a smaller lithium atom number at the end. This indicates that also lithium atoms at high energies with a small angular momentum are cooled by rubidium during the cooling process and not only atoms in the low energy tail. In the experiments with degenerate Fermi-Fermi mixtures, the total heat load of lithium is fine-tuned by varying the initial frequency part (see dashed line in Fig. 5.6 (b)).

Thirdly, cooling of ^{6}Li is assisted by ^{40}K at the end of the sympathetic cooling process. ^{40}K acts here as a catalytic cooling agent and the discussion of can be found below (see Sec. 5.2.3.3).

5.2.3 Quantum degenerate Bose-Fermi-Fermi mixture

Here I present successful sympathetic cooling of the two-species Fermi-Fermi mixture into the quantum degenerate regime. Furthermore, by fine-tuning the starting conditions, a quantum degenerate Fermi-Fermi-Bose mixture can be realized.

In the following, the systematic uncertainty of the presented atom number is conservatively estimated to be below 50 %. The uncertainty for the temperature is lower than $\pm 0.1\, T_F$ for the temperature range $0.2 - 0.5\, T_F$. The absorption images are taken along the z-axis. To increase the signal to noise ratio of these images, the deduced optical column density is projected along the x- or y-axis. The optical line densities are then fitted by a fully physical model function. The bosonic rubidium cloud is fitted by an integrated two-component fit function, which is the sum of a Bose density distribution and a Thomas-Fermi density distribution. The fermionic clouds are fitted by an integrated Fermi-Dirac density distribution.

5. Quantum degeneracy

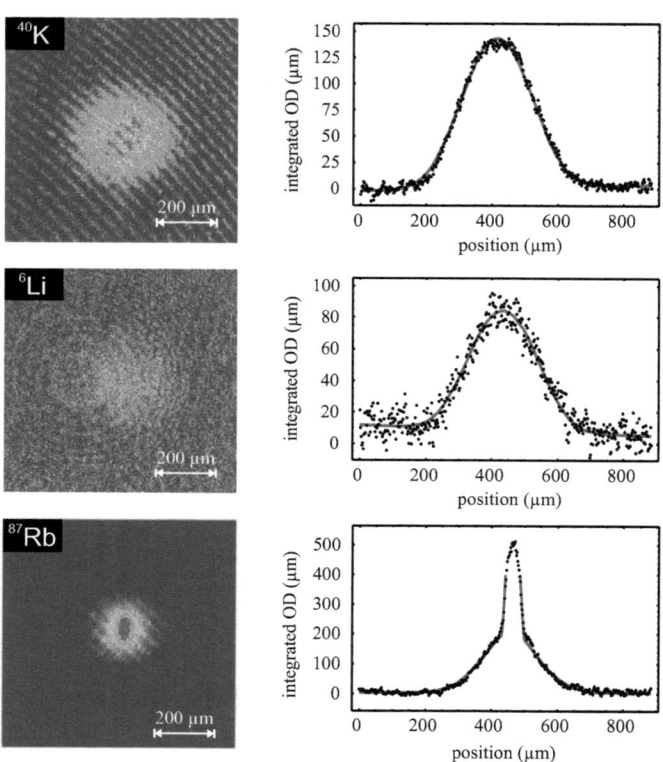

Figure 5.7: Absorption images of the quantum degenerate Fermi-Fermi-Bose mixture. The time-of-flight is 15 ms for potassium, 4 ms for lithium and 20 ms for rubidium.

5.2 Cooling into quantum degeneracy

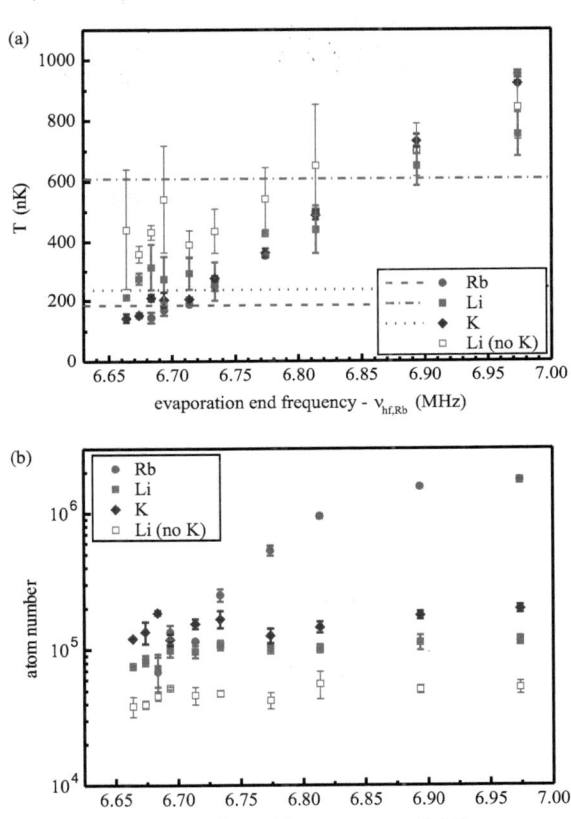

Figure 5.8: (a) Temperatures and (b) atom numbers of the Fermi-Fermi-Bose mixture as a function of the final frequency of the evaporation ramp. The graphs show the last part of the sympathetic cooling process. The horizontal lines in (a) indicate the critical temperature T_c for rubidium and half the Fermi temperature T_F for the fermions. With typical atom numbers of 1.5×10^5 for rubidium and 10^5 for the fermions, the temperatures are $T_c = 190\,\text{nK}$, $T_F^{\text{Li}} = 1.2\,\mu\text{K}$ and $T_F^{K} = 470\,\text{nK}$.

69

5. QUANTUM DEGENERACY

5.2.3.1 Quantum degenerate Fermi-Fermi mixture

With the above mentioned techniques, stretched MW evaporation sequence and removal of high energetic ^{6}Li atoms, it is possible to create a quantum degenerate two-species Fermi mixture.

After 63 s of evaporation, ^{87}Rb is fully evaporated by reducing the MW evaporation frequency below the trap bottom. With optimized parameters, e.g. maximal MOT loading, a mixture of typically $N_{\mathrm{Li}} = 1.8 \times 10^5$ lithium atoms at a temperature of $T/T_{\mathrm{F}} = 0.34$ and $N_{\mathrm{K}} = 1.8 \times 10^5$ potassium atoms at $T/T_{\mathrm{F}} = 0.40$ is realized. These values represent optimal starting conditions for further experiments on a quantum degenerate two-species Fermi mixture.

5.2.3.2 Quantum degenerate Fermi-Fermi-Bose mixture

In order to realize a quantum degenerate Fermi-Fermi-Bose mixture, the former experimental sequence has to be fine-tuned. Since no BEC could be observed, the total heat load on rubidium has be minimized. In the MOT lithium is less loaded by changing the detuning of the lithium MOT beams. This ensures that lithium is always loaded into steady-state, which contribute to an increased stability in the atom number. Furthermore, the background vapor pressure has to be improved by adjusting the dispenser currents.

In Fig. 5.7 a typical Fermi-Fermi-Bose mixture in quantum degeneracy is shown. The expansion times are 15 ms for ^{40}K, 4 ms for ^{6}Li and 20 ms for ^{87}Rb. ^{6}Li expands fastest due to its small mass and this gives an almost comparable cloud size of ^{6}Li and ^{40}K, even at the same temperature. The physical fits give $N_{\mathrm{K}} = 1.3 \times 10^5$ potassium atoms at $T_{\mathrm{K}} = 184\,\mathrm{nK} = 0.35\,T_{\mathrm{F}}^{\mathrm{K}}$ and $N_{\mathrm{Li}} = 0.9 \times 10^5$ lithium atoms at $T_{\mathrm{Li}} = 313\,\mathrm{nK} = 0.3\,T_{\mathrm{F}}^{\mathrm{Li}}$. For rubidium the two-component fit gives $N_{\mathrm{thermal,Rb}} = 1.5 \times 10^5$ atoms in the thermal component and $N_{0,\mathrm{Rb}} = 1 \times 10^5$ in the BEC component at a temperature of $T_{\mathrm{Rb}} = 189\,\mathrm{nK}$.

The cooling process of three-species mixture is further investigated. For this purpose the last part part of the evaporation sequence, above and below quantum degeneracy, is analyzed (see Fig. 5.8). Fig. 5.8 shows the temperature and atom number of the three species as a function of the final frequency of the evaporation ramp. As the three clouds are cooled below the horizontal lines the quantum degenerate regimes are entered. For the two lowest final frequencies the ^{87}Rb cloud can either be not detected or is too small to be analyzed by a least square fit. The curves indicate that ^{87}Rb and ^{40}K are well thermalized throughout the entire cooling. On the other hand, ^{6}Li is higher in temperature than the other two species for lowest MW end frequencies and thus it is not fully thermalized. The fermionic atom numbers are almost constant with $\sim 10^5$ atoms over the presented frequency range and not affected by the MW frequency.

State-pureness of the three species, which is important for further experiments, is checked by Stern-Gerlach separation (see Sec. 3.5) and is evaluated to be better than 97 %.

5.2.3.3 Catalytic cooling

To investigate the cooling process of ^{6}Li further, the temperature and atom number of ^{6}Li is either taken in the presence of ^{87}Rb and ^{40}K or in the presence of ^{87}Rb only (see Fig. 5.8). The curves clearly show that ^{6}Li is cooled more efficiently in the presence of ^{40}K, which is further supported by the increased atom number of ^{6}Li. In the three-species mixture the ^{6}Li atom number is about 10^5, whereas in the ^{6}Li-^{87}Rb mixture the atom number is a factor of 2 smaller. The temperature and atom number dependency imply a ^{6}Li-^{40}K thermalization rate that is comparable or even larger than the ^{6}Li-^{87}Rb thermalization rate. The observation is supported by a recent determination of the scattering length from Feshbach spectroscopy data (Wille et al., 2008), which gives $+63.5(1)\,a_0$ for the triplet scattering length between ^{6}Li and ^{40}K. This gives a scattering cross section that is factor of 10 larger for ^{6}Li-^{40}K than for ^{6}Li-^{87}Rb. A detailed numerical description of the thermalization rate has to include the density overlap of the atomic clouds and, in addition, quantum statistical effects. In conclusion, the increased efficiency of sympathetic cooling lithium by the presence of potassium can be understood as a catalytic cooling process: The potassium cloud acts like a catalytic cooling agent for lithium.

5.3 Conclusions

In conclusion, sympathetic cooling of two species is demonstrated for the first time. Moreover, the first quantum degenerate two-species Fermi-Fermi mixture is realized, which coexists with a Bose-Einstein condensate. The mass difference of lithium and potassium and the different internal structure are new degrees of freedom for the study of fermionic quantum gases.

To achieve quantum degeneracy, different experimental challenges and difficulties of the three species with very different initial temperatures, scattering cross sections and masses were overcome. A combined cMOT/dMOT phase, a careful state cleaning process during evaporation, selective evaporation of rubidium and selective removing of high energetic lithium atoms turn out to be essential for the successful realization of a quantum degenerate Fermi-Fermi-Bose mixture. Furthermore, at the end of the sympathetic cooling process, the cooling efficiency of lithium is improved by catalytic cooling through potassium.

The large atom number of the quantum degenerate Fermi-Fermi-Bose mixture is promising for further experiments. The experimental system allows to explore Fermi-Bose or Fermi-Fermi mixtures. Due to different optical transitions, the mixture offers the convenient way to apply component-selective trapping potentials. Further projects that seem especially interesting here will be in the vicinity of a magnetic Feshbach resonance that change the interaction strength. This should allow to explore the mass difference on the superfluid phase and to create heteronuclear molecules at ultracold temperatures.

5. Quantum degeneracy

Chapter 6

Ultracold heteronuclear Fermi-Fermi molecules

This chapter presents the first creation of ultracold bosonic heteronuclear molecules of two fermionic species, ^{6}Li and ^{40}K, by a magnetic field sweep across an interspecies s-wave Feshbach resonance. This allows to associate up to 4×10^4 molecules with high efficiencies of up to $50\,\%$. Using direct imaging of the molecules, a lifetime increase of the molecules close to resonance of more than $100\,\mathrm{ms}$ in the molecule-atom mixture stored in a harmonic trap is observed.

The chapter is organized as follows: First, the crossed ODT is characterized and the loading sequence of the atoms into the ODT is described in Sec. 6.1. After that, magnetic field calibration of the Feshbach coils and state preparation of ^{6}Li and ^{40}K to the desired states are explained in Sec. 6.2 and Sec. 6.3, respectively. The discussion is followed by identifying Feshbach resonances in Sec. 6.4. Finally, measurements on heteronuclear Fermi-Fermi molecules are presented in Sec. 6.5.

Parts of this chapter are published in:
A.-C. Voigt, M. Taglieber, L. Costa, T. Aoki, W. Wieser, T. W. Hänsch, and K. Dieckmann, "Ultracold Heteronuclear Fermi-Fermi Molecules," Phys. Rev. Lett. **102**, 020405 (2009).

6.1 The optical dipole trap

For the exploration of Feshbach resonances, the Fermi-Fermi mixture is loaded into an optical dipole trap (ODT), which is described in the following. In addition, trap frequencies and heating in the ODT are characterized for quantitative analysis.

6.1.1 Loading into the ODT

The loading scheme from the QUIC trap into the crossed ODT is illustrated in Fig. 6.1. In the QUIC trap a quantum degenerate Fermi-Fermi mixture is created by sympathetic cooling with ^{87}Rb. Lowest temperatures of the fermions are achieved by full evaporation of ^{87}Rb. Before the atoms are loaded into the ODT, remaining ^{87}Rb atoms are removed by a resonant light pulse of $1\,\mathrm{ms}$ length and an intensity of $0.35\,I_\mathrm{sat}$. Additional repumping light during the light pulse brings the rubidium atoms back into the cycling transition, and therefore enhances the efficiency of the cleaning.

6. Ultracold heteronuclear Fermi-Fermi molecules

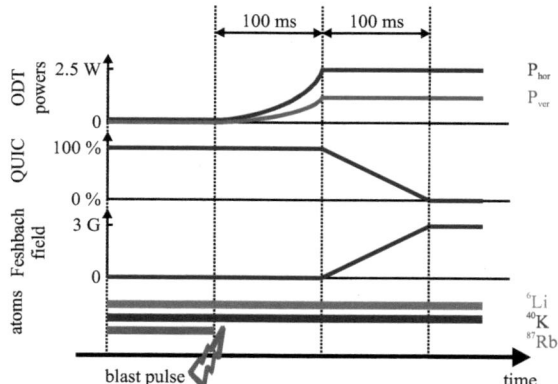

Figure 6.1: Loading scheme of the optical dipole trap. After sympathetic cooling the Fermi-Fermi mixture into quantum degeneracy, remaining ^{87}Rb fractions are removed by a resonant light pulse. Then the horizontal and vertical ODT beams are adiabatically increased to the final value within 100 ms. After that the current of the QUIC trap is linearly decreased to zero within 100 ms.

Then the intensities of the horizontal and vertical ODT beams are adiabatically increased to the final value. For the largest trap frequencies, used in the following, the horizontal and vertical beams are ramped up to 2.44 W and 1.21 W within 100 ms. After that the current of the QUIC trap is linearly decreased to zero within 100 ms. To maintain the magnetic quantization axis, the magnetic Feshbach field is simultaneously increased to 3.2 G.

With this loading scheme a Fermi-Fermi mixture with high atom numbers of $2-3 \times 10^5$ and temperatures of $0.3 - 0.5\,T_F$ for ^{6}Li and ^{40}K can be achieved. These atom numbers are higher than the ones presented in Cha. 5 due to careful choice of the experimental parameters.

6.1.2 Characterization of the ODT

The ODT in crossed beam configuration is fully characterized by the exact knowledge of the two beams waists and the corresponding beam powers. These parameters can be obtained from frequency measurements at different beam powers. The ODT frequencies are measured by parametric heating of the cloud or by exciting the center-of-mass dipole oscillation. A description of the parametric heating method can be found in the diploma thesis (Wieser, 2006). In the following, the oscillation method is exemplarily described for two different beam intensities of the crossed ODT. The atomic cloud is excited by a short magnetic field gradient along one of the axis. After excitation and a variable holding time in the ODT, the displacement of the cloud is measured for a fixed time-of-flight.

Fig. 6.2 shows the observed center-of-mass dipole oscillation of ^{40}K in the z-direction. In Fig. 6.2 (a) the time-of-flight is 6 ms and $P_{hor} = 2.44$ W is the beam power of the horizontal beam and $P_{ver} = 1.21$ W is the power of the vertical beam. In Fig. 6.2 (b) the time-of-flight is 8 ms and the powers are $P_{hor} = 187.9$ mW and $P_{ver} = 95.8$ mW. The influence of the horizontal

6.2 Magnetic field calibration

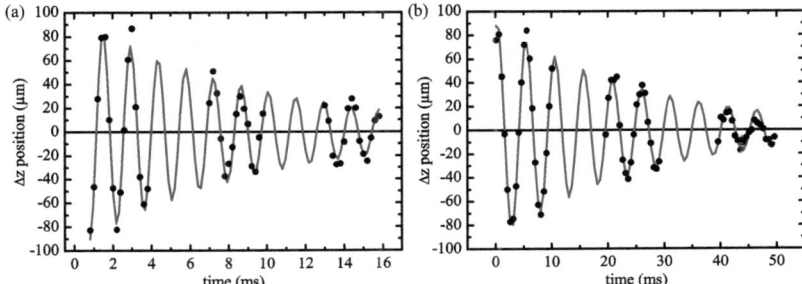

Figure 6.2: Trap frequency measurement of the optical dipole trap along the z-direction for two different trap settings. The clouds are excited by a short magnetic field gradient and the center-of-mass dipole oscillation is monitored. (a) The time-of-flight is 6 ms and the beam powers are $P_{\text{hor}} = 2.44$ W and $P_{ver} = 1.21$ W. (b) The time-of-flight is 8 ms and the powers are $P_{\text{hor}} = 187.9$ mW and $P_{ver} = 95.8$ mW.

beam on the trap frequency in z-direction is smaller than 6 mHz and 2 mHz for case (a) and (b), respectively. The duration of the magnetic field gradient is chosen such that the excursion is well in the harmonic approximation, which is below 10 % deviation from the harmonic approximation in both cases. Additional damping might occur due to slight misalignment of the crossed laser beams, and lead to further damping. The trap frequency $\omega_z^0 = 2\pi\nu_z^0$ is determined by fitting the data with an exponentially damped sinusoidal function

$$\Delta z(t) = A \cdot e^{-\gamma t} \cdot \cos(\omega_z t + \phi) + \text{const.}, \text{ with } \omega_z^0 = \left(\omega_z^2 + \gamma_z^2\right)^{1/2}. \tag{6.1}$$

This yields $\nu_z^0 = 698\,(1)$ Hz for the data in Fig. 6.2 (a) and $\nu_z^0 = 194.8\,(3)$ Hz for the data in Fig. 6.2 (b).

In combination with measurements of the other axes the parameters of the ODT are fully characterized. From this the frequencies for the other two species are calculated.

Due off-resonant scattering of the trap light the clouds are heated up (see Sec. 3.4.6). The heating rate is measured with ^{6}Li for the beam powers as in Fig. 6.2 (a). At 20.6 G ^{6}Li is transferred into the energetically lowest state $|1/2, 1/2\rangle$ to avoid dipolar relaxation. The heating rate is measured to $0.22\,(5)\,\mu$K/s and is in good agreement with the theoretical value of $0.167\,\mu$K/s. This shows that heating due to technical noise is small and demonstrates the stability of the optical mounting and the functionality of the intensity stabilization circuit.

6.2 Magnetic field calibration

Precise control of the magnetic Feshbach field is essential for the exploration of narrow Feshbach resonances. In the following, the magnetic field calibration is discussed.

After installing the Feshbach coils they are initialized by repeatedly switching to a comparatively high magnetic field of several hundred Gauss. This sequence brings the coil assembly into a mechanical equilibrium position, and magnetizes the vacuum chamber and the environment to a reproducible value.

6. Ultracold heteronuclear Fermi-Fermi molecules

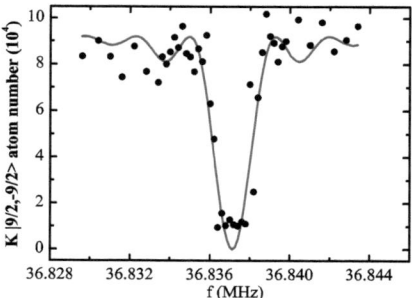

Figure 6.3: Calibration of the magnetic Feshbach field is performed with ^{40}K by driving the $|9/2, -9/2\rangle \rightarrow |9/2, -7/2\rangle$ transition. The graph shows a typical spectrum at 155 G, where the remaining atom number in the $|9/2, -9/2\rangle$ state is recorded for various frequencies in high-field imaging. The curve represents a theoretical fit to the data.

The calibration of the Feshbach coils is performed by driving transitions between different hyperfine states or between Zeeman states within one hyperfine manifold. From the measured transition frequencies the magnetic field can be calculated by the well-known Breit-Rabi formula (Breit and Rabi, 1931).

The first coarse field calibration is done with ^{87}Rb on the MW transition from $|2, 2\rangle \rightarrow |1, 1\rangle$ for several magnetic fields. This transition frequency has a magnetic field sensitivity of 2.1 MHz/G at 0 G, which only slightly increases to higher magnetic fields. The rubidium calibration yields an overall field uncertainty of ± 12 mG from 0 G to 300 G, which is the range of the interesting Feshbach resonances between ^{6}Li and ^{40}K.

The fine calibration is made with the fermionic atom ^{40}K at relevant magnetic field positions, where Feshbach resonances are explored. Spin polarized fermionic atoms have the benefit that the clock shift is absent for transitions at low temperatures due to the Pauli exclusion principle (Gupta et al., 2003). For the calibration measurement a pure ^{40}K$|9/2, -9/2\rangle$ state is prepared that is then transferred by a π-pulse to the $|9/2, -7/2\rangle$ state. The field sensitivity of this transition is 311 kHz/G at 0 G and 180 kHz/G at 155 G. Fig. 6.3 shows a typical calibration spectrum at 155 G, where the remaining atom number in the $|9/2, -9/2\rangle$ state is recorded for various frequencies in high-field imaging. The RF transition uses a fixed square pulse of 400 µs for the π-pulse. The condition for this π-pulse is determined by measuring the transferred atoms as a function of the pulse duration at resonance, which also gives the Rabi frequency $\chi = 2\pi \times 1250$ Hz. To reduce the influence of external magnetic fields on the calibration, the data is taken in the time slot, where no subways run (see Sec. 3.8), and the RF pulse is triggered on the 50 Hz power line signal. The spectrum is fitted with the transition probability for a two-level system, which is given by

$$P_{1 \rightarrow 2} = \frac{\chi^2}{\chi^2 + \Delta^2} \sin^2\left(\frac{1}{2}\sqrt{\chi^2 + \Delta^2}\, T\right) , \text{ with } \Delta = \omega_{12} - \omega . \tag{6.2}$$

In this equation, ω_{12} is the angular transition frequency at resonance, ω is the angular frequency of the driving field and T is the duration of the RF pulse. The fit gives a field uncertainty of

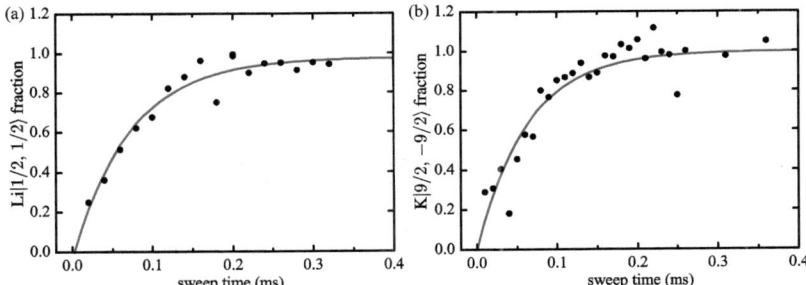

Figure 6.4: State preparation of (a) lithium and (b) potassium by an adiabatic rapid passage. The graphs show the atomic fraction of the final state vs. the sweep time for a fixed frequency span. The curves represent fits with the Landau-Zener formula.

only ±0.3 mG. The excellent agreement between the fit and the data shows that magnetic field inhomogeneities are smaller than the power broadening due to the RF signal. The half width at half maximum of the spectrum is 998 Hz, which corresponds to 5.5 mG. Repeated calibration showed that the drift of the magnetic field over several ones was smaller than 4.6 mG. The overall magnetic field uncertainty, which includes magnetic field inhomogeneities (see Sec. 3.7), external field fluctuations and long-term drifts (see Sec. 3.8), is estimated to be smaller than 7 mG.

6.3 State preparation

For ^{40}K and ^{6}Li several Feshbach resonances in the lowest hyperfine states are reported (Wille et al., 2008). In the following, stability criteria of possible spin mixtures are discussed. Moreover, to explore these Feshbach resonances in detail, the state preparation scheme for ^{40}K and ^{6}Li is shown.

The ODT offers to trap all possible spin states of one atomic species and widens the number of stable spin combinations for atomic mixtures (see also Sec. 5.1.2). In the Fermi-Fermi mixture the spin states that are energetically stable against spin-exchange collisions are

$$^{40}\text{K} \, |9/2, -9/2\rangle \quad \& \quad ^{6}\text{Li} \, |1/2, m_\text{F}\rangle \quad (6.3)$$
$$^{40}\text{K} \, |9/2, -9/2\rangle \quad \& \quad ^{6}\text{Li} \, |3/2, -3/2\rangle \quad (6.4)$$
$$^{40}\text{K} \, |9/2, m_\text{F}\rangle \quad \& \quad ^{6}\text{Li} \, |1/2, 1/2\rangle \, , \quad (6.5)$$

where m_F stays for any possible m_F-level of the respective hyperfine manifold. These combinations are a special feature of the inverted hyperfine structure of ^{40}K and allows to explore a larger set of stable spin combinations.

The state preparation of ^{6}Li and ^{40}K into the desired states is carried out by an adiabatic rapid passage (ARP) (Bloch, 1946; Camparo and Frueholz, 1984; Martin et al., 1988). An electromagnetic field couples two atomic spin states and the frequency of this field is adiabatically swept over the resonance. This procedure coherently transfers one state to another one. The advantage of an ARP is its insensitivity to frequency and power fluctuations of the electromagnetic field and to magnetic field fluctuations.

The state preparation is performed after loading ^{6}Li and ^{40}K into the ODT. The magnetic field is increased to 20.6 G and the atomic states before the ARP are ^{6}Li $|3/2, 3/2\rangle$ and ^{40}K $|9/2, 9/2\rangle$. At this magnetic field the m_F levels are shifted due to the linear and quadratic Zeeman splitting. The quadratic term allows to address each m_F level within one hyperfine manifold individually by selecting the frequency range of the sweep over the transitions into the desired states.

To transfer ^{6}Li from the $|3/2, 3/2\rangle$ to the absolute ground state $|1/2, 1/2\rangle$, an RF field is swept over the resonance from 270.016 MHz to 269.392 MHz. For this two-level system the transition probability $P_{i \to f}$ to the final state is given by the Landau-Zener formula (Zener, 1932; Rubbmark et al., 1981)

$$P_{i \to f} = 1 - \exp(-2\pi\Gamma) \text{ , with } \Gamma = \frac{\Omega^2}{4\dot\omega}. \tag{6.6}$$

Here, Ω is the Rabi frequency and $\dot\omega$ is the frequency sweep speed. The transition is optimized by varying $\dot\omega$ for a given RF power. This is illustrated in Fig. 6.4 (a), which shows the normalized atom number in $|1/2, 1/2\rangle$ vs. time duration for the fixed frequency span mentioned above. The atom numbers in the $|3/2, 3/2\rangle$ and $|1/2, 1/2\rangle$ state are deduced with a state selective Stern-Gerlach method (see. Sec. 3.5.3). Imaging is done at zero magnetic field on the optical $|2S_{1/2}, F = 3/2\rangle \to |2P_{3/2}, F' = 5/2\rangle$ transition. Additional repumping light is used perpendicular to the imaging direction. The curve clearly shows the behavior of a Landau-Zener transition. The fit gives a characteristic $1/e$ time of $\tau = 72\,(13)$ µs, corresponding to a Rabi frequency of $\Omega = 2\pi \times 30(2)$ kHz. To transfer all atoms to the $|1/2, 1/2\rangle$ state, a total time of 7τ is used, which corresponds to a theoretical transition probability of $> 99.9\,\%$.

To prepare the $|9/2, -9/2\rangle$ state of ^{40}K, an RF field is swept from 7.088 MHz to 5.775 MHz. This transfers the $|9/2, 9/2\rangle$ state over all m_F levels to the final $|9/2, -9/2\rangle$ state. As in the case of ^{6}Li, the duration of the transition is optimized for a fixed RF power. Fig. 6.4 (b) shows the normalized number of atoms in the $|9/2, -9/2\rangle$ state. The atom numbers in each spin state are imaged after a Stern-Gerlach separation. Here, the characteristic $1/e$ time is $\tau = 63\,(10)$ µs. This corresponds to an effective Rabi frequency over the involved transitions of $\Omega = 2\pi \times 18.3(1.4)$ kHz. As before, a total time of 7τ is used to transfer the atoms into the $|9/2, -9/2\rangle$ state. In the following sections, a $|9/2, -5/2\rangle$ state is used to explore a specific Feshbach resonance. This state is prepared after increasing the magnetic field to the desired value near the Feshbach resonance. A second ARP from $|9/2, -9/2\rangle$ to $|9/2, -5/2\rangle$ prepares the state.

In both cases, the remarkable short transfer times are faster than the inverse trapping frequencies and collision times. This enables transitions with negligible dephasing due to collisions.

Due to stability arguments, it is favorable to transfer first ^{6}Li to the absolute ground state and then ^{40}K (see stability discussion above).

6.4 Identifying Feshbach resonances

Several s- and p-wave Feshbach resonances between ^{6}Li and ^{40}K are identified by a non-degenerate mixture through loss measurements at the Innsbruck group (Wille et al., 2008). By means of the asymptotic bound state model the measured loss features have been assigned

6.4 Identifying Feshbach resonances

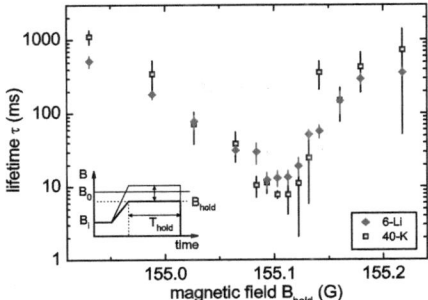

Figure 6.5: Lifetime of the ^{6}Li-^{40}K mixture as a function of the magnetic field in the vicinity of a Feshbach resonance between ^{6}Li $|1/2, 1/2\rangle$ and ^{40}K $|9/2, -5/2\rangle$.

to s- and p- wave resonances. This has been further confirmed by a full coupled channel calculation (Stoof et al., 1988) that, moreover, yields the singlet and triplet background scattering lengths between the two fermions and the widths of the Feshbach resonances.

According to this calculation all resonances are rather narrow and expected to be closed-channel dominated. In the following sections, a s-wave Feshbach resonance with a comparative large width of 0.81 G between ^{6}Li $|1/2, 1/2\rangle$ and ^{40}K $|9/2, -5/2\rangle$ at 155.1 G is analyzed. Here, the coupling energy scale is $E_0 \sim k_B \times 1\,\mu$K and is comparable to the experimental realized Fermi energies. Therefore, the closed molecular channel affects the many-body physics (see Sec. 2.2.4).

6.4.1 Loss measurement

In the following, the lifetime of the mixture close to the Feshbach resonance as a function of the magnetic field is investigated (see Fig. 6.5).

For this measurement, ^{6}Li and ^{40}K are loaded into the nearly isotropic ODT with mean trapping frequencies of $\bar{\nu}_{Li} = 1245(30)$ Hz for ^{6}Li and $\bar{\nu}_K = 725(20)$ for ^{40}K. In this trap, the typical atom numbers of $N_{Li} \approx N_K \approx 1 \times 10^5$ and temperatures of $T_{Li} = 0.4\,T_F^{Li}$ and $T_K = 0.6\,T_F^K$ correspond to peak densities of $n_{Li} = 2.0 \times 10^{13}\,\text{cm}^{-3}$ and $n_K = 1.5 \times 10^{14}\,\text{cm}^{-3}$. Starting from the optically trapped ultracold mixture of ^{6}Li and ^{40}K atoms in the absolute ground states, the magnetic bias field is linearly ramped within 30 ms from 20.6 G to a constant value of $B_i = 152.78$ G, i.e. to the molecular side below the Feshbach resonance (see inset of Fig. 6.5). The atoms are then transferred into the hyperfine states ^{40}K $|9/2, -5/2\rangle$ and ^{6}Li $|1/2, 1/2\rangle$ by means of an ARP. In a second linear ramp, the magnetic field is increased within 0.5 ms to a variable value B_{hold}, and kept constant for a duration T_{hold}. It is then rapidly reduced to 1 G, with an initial slope of 820 G/ms. After a subsequent holding time of 5 ms, the atoms are released from the ODT. Finally, the lithium and potassium clouds are detected by absorption imaging after 1 ms and 4 ms of free expansion. In order to determine the lifetime τ of the gas for a given magnetic field B_{hold}, the experiment is repeated for at least eight different holding durations T_{hold} and an exponential decay function is fitted to the obtained atom numbers resulting in an exponential lifetime τ. As shown in Fig. 6.5, the lifetime τ of

6. Ultracold heteronuclear Fermi-Fermi molecules

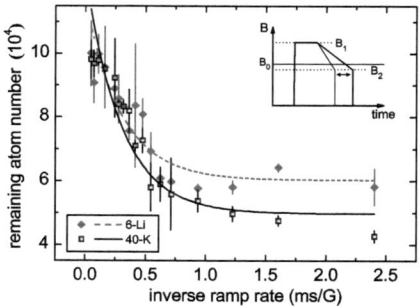

Figure 6.6: Adiabaticity of the molecule conversion process as a function of the inverse sweep rate of the magnetic field strength. The decrease in atom number is attributed to molecule association. The lines represent an exponential fit based on the Landau-Zener theory.

the mixture as a function of B_{hold} decreases by two orders of magnitude to a minimum of 10 ms at 155.10 (5) G. Note that this width is significantly smaller than the one obtained in (Wille et al., 2008) for the same Feshbach resonance but at much higher temperature.

The observed decrease in atom number is attributed to the creation of molecules, which are not detected by the imaging procedure, and to losses of atoms and molecules from the trap due to vibrational relaxation. The observed weak asymmetry of the line shape is in qualitative agreement with predictions for three-body relaxation (D'Incao and Esry, 2006).

6.5 Heteronuclear Fermi-Fermi Molecules

Within this thesis, weakly bound bosonic Feshbach molecules of two different fermionic species are created by an adiabatic magnetic field sweep across the Feshbach resonance (van Abeelen and Verhar, 1999; Timmermans et al., 1999; Mies et al., 2000). In the following, the molecular creation process and lifetime are analyzed. Moreover, a direct detection method of the molecules is presented, which is favorable for the heteronuclear case.

6.5.1 Adiabatic conversion process

In the first measurement, we convert the two-species mixture to heteronuclear molecules by sweeping the magnetic field strength from the atomic to the molecular side of the Feshbach resonance. We study the adiabaticity of the association process as a function of the sweep rate $\dot{B}$ of the magnetic field strength.

For this measurement, the ODT parameters are the same as described in the loss measurement. Now, the ^{6}Li $|1/2, 1/2\rangle$ and ^{40}K $|9/2, -5/2\rangle$ states are prepared at a magnetic field strength of $B_1 = 156.91$ G far on the atomic side of the Feshbach resonance. The magnetic field strength is then linearly ramped with a variable ramp speed to a fixed final value $B_2 = 154.73$ G far on the molecular side of the Feshbach resonance. As before, the magnetic bias field is then rapidly switched off and the remaining atoms are detected after 5 ms holding time in the ODT

6.5 Heteronuclear Fermi-Fermi Molecules

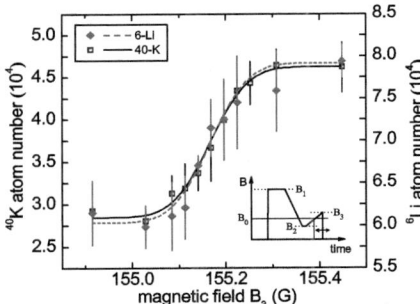

Figure 6.7: Reconversion process of molecules to unbound atoms as a function of the final magnetic field strength. First, molecules are associated by an adiabatic sweep from B_1 to B_2. Then, the molecules are adiabatically reconverted to unbound atoms by a sweep to a variable final value B_3. The lines represent an empirical fit using an error function.

and subsequent free expansion. Since the imaging procedure at low magnetic field detects only free atoms, the creation of molecules shows up in Fig. 6.6 as a decrease in the observed atom number. In this way we typically detect 4×10^4 molecules with a conversion efficiency close to 50 %. To my knowledge, this is the largest heteronuclear molecule number, reported so far.

According to a Landau-Zener behavior of the conversion process of atoms into molecules (Chwedeńczuk et al., 2004; Hodby et al., 2005; Köhler et al., 2006) an exponential function is fitted to the remaining atom numbers. We obtain a characteristic inverse sweep rate of $1/\dot{B}_{\mathrm{ad}} = 0.30(5)$ ms/G for adiabatic association of atoms into molecules. Our sweep rate $\dot{B}_{\mathrm{ad}}$ is smaller than in fermionic experiments with broad resonances (Regal et al., 2003; Schunck et al., 2005), as one would expect for closed-channel dominated resonances due to a small coupling matrix element.

6.5.2 Reconversion process

Next, it is shown that the observed decrease in atom number can indeed be attributed to the creation of molecules (see Fig. 6.7).

For this purpose, the atoms are prepared in the ODT in the states $^6\mathrm{Li}\,|1/2,\,1/2\rangle$ and $^{40}\mathrm{K}\,|9/2,\,-5/2\rangle$ at the magnetic field strength B_1 as in the previous measurement. Free atoms are converted into weakly bound molecules by ramping the magnetic field strength to the value B_2 on the molecular side of the Feshbach resonance. The field is linearly ramped with a constant rate of 1 G/ms, which slower than the characteristic sweep rate $\dot{B}_{\mathrm{ad}}$. After 200 ms, the magnetic field strength is linearly ramped back with the same sweep rate to a variable final value B_3. As before, the magnetic field is then rapidly switched off and the atoms are detected after 5 ms holding time in the ODT and subsequent free expansion. Fig. 6.7 shows the lithium and potassium atom numbers as a function of the final magnetic field B_3.

Towards higher final magnetic fields B_3, molecules are dissociated back into atoms resulting in atom count for both species that is increased by $1.8(3) \times 10^4$. Note that this increase does

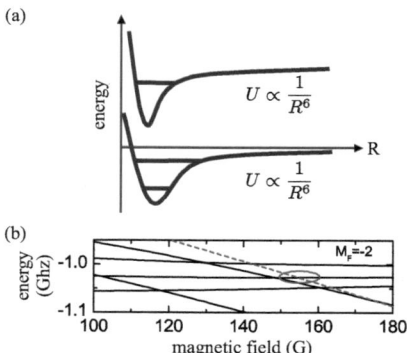

Figure 6.8: (a) Heteronuclear molecules exhibit the same asymptotic behavior of the ground and the first excited states. Both scale as $1/R^6$ with the internuclear separation R. In the homonuclear case the ground state vary as $1/R^6$ and the first excited state vary as $1/R^3$. (b) Energies of molecular states (solid lines) and free atoms (dashed line): The molecular state has almost vanishing magnetic moment at the Feshbach resonance (encircled region).

not reflect the total number of molecules created, as loss of the molecules occurs during holding and the dissociation sweep (see Fig. 6.11). Furthermore, the same scaling of the atom numbers for ^{6}Li and ^{40}K in Fig. 6.7 should be highlighted. The increase of the atom signal is the same for both species and reflects the excellent atom calibration, which is independently done for each of them. By an empirical fit to the number of detected atoms, the reconversion process is found to be centered at 155.17(8) G, where the uncertainty represents the 10 % and 90 % levels.

In addition, molecules at another, narrower Feshbach resonance between the states ^{6}Li $|1/2, 1/2\rangle$ and ^{40}K $|9/2, 9/2\rangle$ are created, which is located at 168.6(2) G. For this resonance, up to $2.0(5) \times 10^4$ molecules are associated by the magnetic sweep technique.

6.5.3 Direct molecule detection and molecule purification

The molecule detection by adiabatically sweeping over the Feshbach resonance, as presented above, is an indirect method. Here, the real molecule signal can be inferred by taking the difference between the atom signal after conversion and reconversion process and including losses during reconversion. In the following, I present a purification technique that allows direct and simultaneous detection of the molecule and atom signal and greatly increases the detection sensitivity.

6.5.3.1 Principle

Feshbach molecules with small binding energies can be detected by direct absorption imaging. The resonant transition is only slightly detuned compared to the unbound case. The first scattered photon presumably dissociate the weakly bound molecules into the free atoms, of which the subsequent photons are scattered. The heteronuclear case is favored compared to

6.5 Heteronuclear Fermi-Fermi Molecules

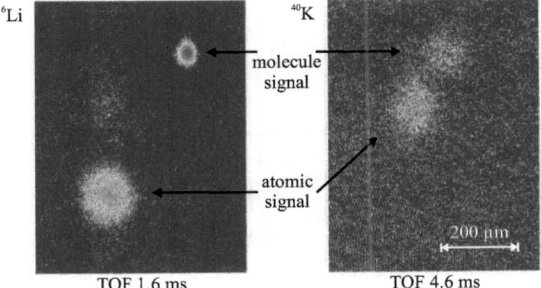

Figure 6.9: Direct simultaneous absorption imaging of the molecular and atomic cloud after separation by a Stern-Gerlach force during time-of-flight (TOF). The absorption signals for ^{6}Li and ^{40}K differ due to different Clebsch-Gordon coefficients.

the homonuclear case by the fact that the ground and the first excited states share the same asymptotic behavior (Zirbel et al., 2008b; Ospelkaus et al., 2008). For weakly bound heteronuclear molecules both states vary with the internuclear separation R as $1/R^6$, which is illustrated in Fig. 6.8. For homonuclear molecules the ground state vary as $1/R^6$ and due to resonant dipole-dipole interaction the first excited state has a long-range scaling as $1/R^3$. This allows to detect heteronuclear Feshbach molecules over a wider magnetic field range than homonuclear ones.

Purification of the molecules from the unbound atoms can be obtained by spatially separating the atoms from the molecular cloud by applying a Stern-Gerlach force before imaging. For this purpose a difference of the magnetic moments of the molecules, $\mu(^6$Li^{40}K$)$, and the free atoms, $\mu(^6$Li$)$ and $\mu(^{40}$K$)$, is exploited. This difference can be inferred from the magnetic field dependence of the molecular bound state energy as compared to the energies of the two free atoms, as shown in Fig. 6.8 (b). The dependence of the molecular energy is calculated with the asymptotic bound state model (Wille et al., 2008) for the manifold with a total projection quantum number $M_F = -2$, which is relevant for the states ^{6}Li $|1/2, 1/2\rangle$ and ^{40}K $|9/2, -5/2\rangle$.

The combination of direct imaging and purification allows to sensitively detect the molecular and atomic component of one species with only one absorption image. Furthermore, no additional molecular dissociation sequence by a magnetic field sweep or by a RF pulse is needed.

6.5.3.2 Demonstration of direct imaging

For the following series of measurements, the ODT is weakened to work at lower densities and colder temperatures. The mean trapping frequencies are $\bar{\nu}_{Li} = 347(8)$ Hz for lithium and $\bar{\nu}_K = 203(5)$ Hz for potassium.

To demonstrate the direct detection scheme, molecules are prepared at 155.03 G. Then the ODT is switched off and a magnetic field gradient pulse with an average strength of 167 G/cm is applied for 570 µs. After a total time of free expansion of 1.6 ms for ^{6}Li and 4.6 ms for ^{40}K, the atomic and molecular clouds are simultaneously imaged using a closed optical transition at the constant high magnetic Feshbach field. The resulting images presented in Fig. 6.9 directly show that the molecular cloud is clearly separated from the atomic cloud. The different masses of the

6. Ultracold heteronuclear Fermi-Fermi molecules

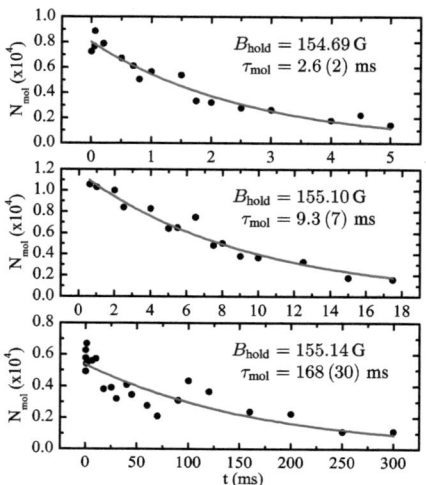

Figure 6.10: Lifetime measurement of the molecules in the molecule-atom mixture. Three measurements at different magnetic field strengths B_{hold} in the vicinity of the Feshbach resonance are exemplarily shown. An exponential fit to the data gives the molecular $1/e$ lifetime τ_{mol}.

molecules and atoms result in different expansion velocities of the corresponding clouds, even at equal temperature. The absorption signals for ^{6}Li and ^{40}K differ due to different transition strengths in high-field imaging.

The magnetic moment of the molecules can be determined from measuring the trajectory of the molecules in a variable magnetic field gradient. From this, a negligible magnetic moment of the molecules of $< 0.1\,\mu_B$ is extracted, as expected from the asymptotic bound state model.

6.5.4 Lifetime measurement

With the previous purification and direct detection method the $1/e$ lifetime of the molecules τ_{mol} in the molecule-atom mixture as a function of the magnetic field strength is investigated (see Fig. 6.10 and 6.11).

The heteronuclear molecules are created by an adiabatic linear ramp of the magnetic field strength from B_1 to a variable final value B_{hold}. For each value of B_{hold}, the molecule-atom mixture is held for a variable duration in the weak ODT before the molecular and atomic clouds are released from the trap and detected separately as described in the previous measurement. For this measurement, the peak densities before the magnetic field sweep are $n_{\text{Li}} = 2.9 \times 10^{12}\,\text{cm}^{-3}$ and $n_{\text{K}} = 2.2 \times 10^{13}\,\text{cm}^{-3}$ with temperatures $T_{\text{Li}} = 0.3\,T_F^{\text{Li}}$ and $T_{\text{K}} = 0.4\,T_F^{\text{K}}$. The trapping frequencies are the same as in the previous Sec. 6.5.3.

Three exemplary measurements of the molecular signal at different magnetic fields are shown in Fig. 6.10. The lifetime as a function of the magnetic field strength is plotted in Fig. 6.11. In the molecule-atom mixture, lifetimes τ_{mol} of the heteronuclear molecules of more than

6.6 Discussion & conclusions

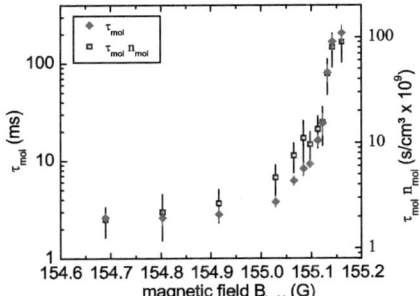

Figure 6.11: Characterization of the molecular $1/e$ lifetime τ_{mol} in the molecule-atom mixture as a function of the magnetic field strength. In addition, $\tau_{\text{mol}} n_{\text{mol}}$ is plotted, where n_{mol} is the initial molecular density, averaged over the cloud. The data shows an increase of τ_{mol} by almost two orders of magnitude for magnetic field strengths close to the resonance compared to the value far away from it on the molecular side.

100 ms are observed. Importantly, it is found that τ_{mol} is increased by almost two orders of magnitude for magnetic field strengths close to the resonance compared to the value far away from the resonance on the molecular side. This observation is consistent with previous results for homonuclear fermionic spin mixtures (Strecker et al., 2003; Cubizolles et al., 2003; Regal et al., 2003; Dieckmann et al., 2002; Regal et al., 2004a). Since due to the molecule creation process the densities of the molecules and atoms depend on B_{hold}, we have to show that the increased lifetime close to the resonance is not only a mere consequence of a smaller molecule density. For this purpose, Fig. 6.11 also shows the product of the lifetime and the initial average density of the molecules, $\tau_{\text{mol}} n_{\text{mol}}$. This corresponds to the inverse constant of a second order decay. Also this parameter is significantly increased close to resonance. Note that such suppression of inelastic decay has been explained on the basis of the Pauli principle for the case of an open-channel dominated Feshbach resonance between fermionic atoms (Petrov et al., 2005). However, the Feshbach resonances used in this work are expected to be closed-channel dominated (Wille et al., 2008).

6.6 Discussion & conclusions

The initial temperatures for ^{6}Li and ^{40}K are smaller than the predicted critical temperature $T_c = 0.52\, T_F$ for conversion into a heteronuclear molecular BEC (Ohashi and Griffin, 2003). While BEC has not been observed, the long molecular lifetime is a good basis to enter this regime by additional cooling. The detection of the bimodal density distribution of the thermal and condensed part might be improved by changing the trap geometry from a nearly isotropic trap to an elongated trap. Possible heating of the molecular cloud could be minimized by carefully removing unbound atoms. Further, a simple estimation based on our experimental parameters and on the width of the Feshbach resonance given in Ref. (Wille et al., 2008) show that the Fermi-Fermi two-species mixture can be completely prepared in the strongly interacting regime ($k_F |a| \geq 1$, where k_F is the Fermi wave vector and a the s-wave scattering length). The magnetic range of this regime is estimated to be $\sim 35\,\mathrm{mG}$ (see also discussion on magnetic field inhomogeneities in Sec. 3.7 and ambient magnetic fields in Sec. 3.8). This, in combination with the long lifetime, may allow us to study many-body physics of a heteronuclear mixture at a closed-channel Feshbach resonance.

A universal Feshbach molecular state implies a negligible closed channel admixture. According to Eq. (2.53) this is fulfilled for a magnetic field control of far below $\sim 10\,\mathrm{mG}$ at the Feshbach resonance. Eliminating ambient magnetic fields and further increasing the long term drifts gives already a magnetic field control below $1\,\mathrm{mG}$ in our experimental setup (see Sec. 6.2). The exact magnetic field range of universal scaling of the molecular binding energy can be precisely measured, for example, by RF spectroscopy (Regal et al., 2003; Chin and Julienne, 2005). The condition $E_0 \gg E_\mathrm{F}$ for universal behavior throughout the entire strongly interacting regime (see Sec. 2.2.4) requires Fermi energies much smaller than $\sim k_\mathrm{B} \times 1\,\mathrm{\mu K}$. This could be achieved with low density ensembles by lowering the particle number or the trapping frequencies into the Hertz regime, while maintaining submilligauss magnetic field control over an increased trapping volume. The relative gravitational sag between the two fermionic species could be compensated by an additional levitation field or by a well-chosen frequency of the trapping light for which both species experience the same trapping frequency (e.g. $799.5\,\mathrm{nm}$ for the $^6\mathrm{Li} - {}^{40}\mathrm{K}$ mixture (Safronova et al., 2006)).

Chapter 7
Conclusions and Outlook

This thesis presents a new and very versatile experimental platform to study quantum degenerate two-species Fermi mixtures. The experimental concept is sympathetic cooling of two fermionic species (^{40}K and ^{6}Li) by a large bosonic gas (^{87}Rb). It is shown that studying a system of three quantum degenerate atomic species is possible for the first time. Reliable operation is guaranteed by careful choice of the experimental components, which is essential to deal with the complexity of such an experimental platform with three species.

The first milestone towards quantum degeneracy is realized by simultaneous trapping ^{87}Rb, ^{40}K and ^{6}Li in a magneto-optical trap ("triple MOT"). This marks the first realization of magneto-optical trapping of two fermionic species and also of three species. The achieved atom numbers and temperatures in combination with the observed losses due to light-assisted collisions, which are at tolerable values, are excellent starting conditions to sympathetically cool the Fermi-Fermi mixture.

Successful sympathetic cooling of ^{6}Li and ^{40}K into quantum degeneracy with large atom numbers is deomonstrated. This presents the first quantum degenerate mixture of two fermionic species. In addition, the first quantum degenerate Fermi-Fermi mixture coexisting with a BEC is realized. It is found that catalytic cooling can enhance the sympathetic cooling efficiency. The two-species Fermi-Fermi mixture adds the mass ratio and the different internal structure as new parameters to fermionic quantum gases and allows to conveniently apply component-selective trapping potentials.

The last presented experimental results demonstrate that a Fermi-Fermi mixture can be used to create ultracold bosonic heteronuclear molecules of two different species for the first time. The production process by an adiabatic magnetic field sweep across an interspecies s-wave Feshbach resonance has high efficiencies of up to 50 %. The associated number of molecules are up to 4×10^4 molecules, which is the largest number of heteronuclear molecules produced so far. Furthermore, close to resonance an increased molecular lifetime of more than 100 ms is observed.

These results open up two interesting directions for future experiments: many-body physics of different masses and quantum gases of dipolar ground state molecules.

The long lifetimes of the molecules at resonance with temperatures close to quantum degeneracy are promising to create a BEC of heteronuclear molecules and to study the BEC-BCS crossover with different masses. Even for unequal masses Cooper pairing is still possible with zero net-momentum of the pairs, if the densities of the two fermionic components are equal. In the BCS limit the critical temperature and order parameter are affected in a nontrivial way

7. Conclusions and Outlook

by the mass ratio (Baranov et al., 2008). The exploration of the nature and existence of superfluidity in the case of unmatched Fermi surfaces, realized by unequal densities of the two fermionic components, is an interesting and fascinating direction for future studies. Here many new quantum phases and exotic pairing mechanisms are expected that might be of particular relevance for high-T_c superconductors (Casalbuoni and Nardulli, 2004) and for analogies to baryonic phases of quantum chromodynamics (Wilczek, 2007; Rapp et al., 2007). The Fulde-Ferrell-Larkin-Ovchinnikov (FFLO) state (Fulde and Ferrell, 1964; Larkin and Ovchinnikov, 1965), for example, has Cooper pairs with a finite net-momentum and a spatially oscillating order parameter. Such a FFLO state and its dependency on the fermionic mass ratio might be experimentally accessible in a one-dimensional optical lattice, e.g. (Orso, 2007; Feiguin and Heidrich-Meisner, 2007; Wang et al., 2009). Moreover, a gas-crystal quantum transition is predicted for large mass ratios (Petrov et al., 2007). Due to the different internal structure of ^{6}Li and ^{40}K, the effective mass ratio could be changed conveniently by an optical lattice.

The successful realization of weakly bound heteronuclear molecules is an excellent starting point to transfer them to the rovibrational ground state and to create polar bosonic molecules with an appreciable, large dipole moment (Aymar and Dulieu, 2005). A highly efficient transfer that keeps the molecular phase space density high could be achieved by a coherent stimulated Raman adiabatic passage (STIRAP) (Bergmann et al., 1998), as demonstrated recently with ^{87}Rb$_2$ (Lang et al., 2008), ^{133}Cs$_2$ (Danzl et al., 2008) and ^{40}K^{87}Rb (Ni et al., 2008). In general, for heteronuclear alkali molecules it is beneficial to work with the STIRAP on the energetically lowest atomic asymptote, which is the ^{40}K-asymptote in our case. The reason are typically repulsive van der Waals and dipole-quadrupole interactions in the excited states (Vadla et al., 1983; Movre and Beuc, 1984; Bussery et al., 1987; Wang and Stwalley, 1998). The bosonic character of the molecules makes it promising to cool them further by evaporative cooling and to create a polar BEC with an anisotropic, long-range interaction (Santos et al., 2000, 2002; Góral and Santos, 2002). In combination with optical lattices many new quantum phases are predicted (Góral et al., 2002; Micheli et al., 2006; Barnett et al., 2006), such as quantum matter states with topological order (Micheli et al., 2006). In addition, the application to quantum information processing (de Mille, 2002) seems to be promising due to effective coupling of the electric dipoles at moderate distances and the relatively easy way of controlling the electric dipole moments by electric fields. Another exciting research direction with ultracold polar molecules is to test fundamental physics. Of particular interest might be to study time variation of fundamental constants (Hudson et al., 2006a; Chin et al., 2009) and violation effects of parity and time-reversal (Kozlov and de Mille, 2002; Hudson et al., 2002).

Appendix A

Natural constants and atomic sources

A.1 Natural constants

quantity	symbol	unit	value
speed of light	c	299 792 458	$\mathrm{m\,s^{-1}}$
Planck's constant	h	$6.6260693\,(11) \times 10^{-34}$	$\mathrm{J\,s}$
electron charge	e	$1.60217653\,(14) \times 10^{-19}$	C
Bohr magneton	μ_B	$9.27400949\,(80) \times 10^{-24}$	$\mathrm{J\,T^{-1}}$
nuclear magneton	μ_N	$5.05078343\,(43) \times 10^{-27}$	$\mathrm{J\,T^{-1}}$
Bohr radius	a_0	$0.5291772108\,(18) \times 10^{-10}$	m
electron mass	m_e	$9.1093826\,(16) \times 10^{-31}$	kg
Boltzmann constant	k_B	$1.3806504\,(24) \times 10^{-23}$	$\mathrm{J\,K^{-1}}$
atomic mass unit	u	$1.660538782\,(83) \times 10^{-27}$	kg

Table A.1: The values of the fundamental physical constants are taken from the CODATA database that can be found at http://physics.nist.gov/cuu/Constants/.

A.2 Atomic properties

symbol	^{6}Li	^{40}K	^{87}Rb	unit	references
η	7.59 (4) %	0.011 7 (1) %	27.83 (2) %		(1)
m	6.015 122 3(5)	39.963 998 67 (29)	86.909 187 35 (27)	u	(2)
$T_{1/2}$	stable	1.25×10^9	4.81×10^{10}	years	(1)
I	1	4	3/2		(1)
g_I	4.476 540 (3)	1.765 490 (34)	−9.951 414 (10)	10^{-4}	(3)
ν_{hf}	228.205 28 (8)	1285.79 (1)	6834.682 610 9 (3)	MHz	(3)
a_S	+38.75	+104.8 (4)	+90.6	a_0	(4)
a_T	−2240	+174 (7)	+98.96	a_0	(4)
τ	27.102	25.7	26.24	ns	(5)
Γ	5.872 4	6.2	6.065	MHz	(6)
λ_2	670.977	766.7	780.246	nm	(7)
I_s	2.541	1.8	1.669	m W cm^{-2}	
v_{rec}	9.9	1.3	0.59	cm/s	
T_{rec}	7089	814	362	nK	
T_{D}	141	149	146	µK	

Table A.2: The most important atomic properties of the atomic species used within this thesis: the relative abundance η of the isotope, the atomic mass m, the nuclear half-lifetime $T_{1/2}$, the nuclear spin I, the g-factor g_I, the ground state hyperfine splitting ν_{hf}, the singlet s-wave scattering length a_S, the triplet s-wave scattering length a_T, the lifetime τ of the excited $P_{3/2}$ state, the corresponding linewidth Γ, the resonance wavelength λ_2 of the D_2-line in vacuum, the saturation intensity $I_s = \pi\,h\,c\,\Gamma/(3\,\lambda_2^3)$, the photon recoil velocity $v_{\text{rec}} = h/(m\,\lambda_2)$, the photon recoil temperature $T_{\text{rec}} = m\,v_{\text{rec}}^2/k_{\text{B}}$, and the Doppler temperature $T_{\text{D}} = \hbar\,\Gamma/(2\,k_{\text{B}})$.

---

[1] (Brookhaven)
[2] (Coursey et al.)
[3] (Arimondo et al., 1977)
[4] (O'Hara, 2000; Loftus et al., 2002; Marte et al., 2002; van Kempen et al., 2002)
[5] (McAlexander et al., 1996; Volz and Schmoranzer, 1996)
[6] (Roati et al., 2001)
[7] (Scherf et al., 1996; Ye et al., 1996; Roati et al., 2001)

quantity	symbol	value	unit	reference
^{6}Li – ^{87}Rb triplet scattering length	$a_{T,\text{LiRb}}$	-19.8	a_0	[8]
^{40}K – ^{87}Rb triplet scattering length	$a_{T,\text{KRb}}$	$-215\,(10)$	a_0	[9]
^{6}Li – ^{40}K triplet scattering length	$a_{T,\text{LiK}}$	$+63.5\,(1)$	a_0	[10]

Table A.3: The triplet interspecies scattering lengths between ^{6}Li, ^{40}K and ^{87}Rb.

---

[8] (Li et al., 2008)
[9] (Ferlaino et al., 2006a,b)
[10] (Wille et al., 2008)

A. Natural constants and atomic sources

Bibliography

van Abeelen, F. A., and B. J. Verhar, "Time-Dependent Feshbach Resonance Scattering and Anomalous Decay of a Na Bose-Einstein Condensate," Phys. Rev. Lett. **83**, 1550 (1999).

Adams, C. S., H. J. Lee, N. Davidson, M. Kasevich, and S. Chu, "Evaporative Cooling in a Crossed Dipole Trap," Phys. Rev. Lett. **74**, 3577 (1995).

Altman, E., E. Demler, and M. D. Lukin, "Probing many-body states of ultracold atoms via noise correlations," Phys. Rev. A **70**, 013603 (2004).

Anderson, M. H., J. R. Ensher, M. R. Matthews, C. W. Wieman, and E. A. Cornell, "Observation of Bose-Einstein Condensation in a Dilute Atomic Vapor," Science **269**, 198 (1995).

Andreev, A. V., V. Gurarie, and L. Radzihovsky, "Nonequilibrium Dynamics and Thermodynamics of a Degenerate Fermi Gas Across a Feshbach Resonance," Phys. Rev. Lett. **93**, 130402 (2004).

Arimondo, E., M. Inguscio, and P. Violino, "Experimental determination of the hyperfine structure in the alkali atoms," Rev. Mod. Phys. **49**, 31 (1977).

Aubin, S., S. Myrskog, M. Extavour, L. Leblanc, D. McKay, A. Stumm, and J. Thywissen, "Rapid sympathetic cooling to Fermi degeneracy on a chip," Nature Physics **2**, 384 (2006).

Aymar, M., and O. Dulieu, "Calculation of accurate permanent dipole moments of the lowest $^{1,3}\Sigma^+$ states of heteronuclear alkali dimers using extended basis sets," J. Chem. Phys. **122**, 204302 (2005).

Bahns, J. T., W. C. Stwalley, and P. L. Gould, "Laser cooling of molecules: A sequential scheme for rotation, translation, and vibration," J. Chem. Phys. **104**, 9689 (1996).

Baker, G. A., "Neutron matter model," Phys. Rev. C **60**, 054311 (1999).

Baranov, M. A., C. Lobo, and G. V. Shlyapnikov, "Superfluid pairing between fermions with unequal masses," Phys. Rev. A **78**, 033620 (2008).

Baranov, M. A., M. S. Mar'enko, V. S. Rychkov, and G. V. Shlyapnikov, "Superfluid pairing in a polarized dipolar Fermi gas," Phys. Rev. A **66**, 013606 (2002).

Bardeen, J., L. N. Cooper, and J. R. Schrieffer, "Theory of superconductivity," Phys. Rev. **108**, 1175 (1957).

Barnett, R., D. Petrov, M. Lukin, and E. Demler, "Quantum Magnetism with Multicomponent Dipolar Molecules in an Optical Lattice," Phys. Rev. Lett. **96**, 190401 (2006).

Bartenstein, M., A. Altmeyer, S. Riedl, S. Jochim, C. Chin, J. H. Denschlag, and R. Grimm, "Collective excitations of a degenerate gas at the BEC-BCS crossover," Phys. Rev. Lett. **92**, 203201 (2004a).

Bartenstein, M., A. Altmeyer, S. Riedl, S. Jochim, C. Chin, J. H. Denschlag, and R. Grimm, "Crossover from a molecular Bose-Einstein condensate to a degenerate Fermi gas," Phys. Rev. Lett. **92**, 120401 (2004b).

Bedaque, P. F., H. Caldas, and G. Rupak, "Phase separation in asymmetrical fermion superfluids," Phys. Rev. Lett. **91**, 247002 (2003).

Bergmann, K., H. Theuer, and B. W. Shore, "Coherent population transfer among quantum states of atoms and molecules," Rev. Mod. Phys. **70**, 1003 (1998).

Bethlem, H. L., G. Berden, and G. Meijer, "Decelerating Neutral Dipolar Molecules," Phys. Rev. Lett. **83**, 1558 (1999).

Bjorklund, G., M. Levenson, W. Lenth, and C. Ortiz, "Frequency modulation (FM) spectroscopy," Appl. Phys. B **32**, 145 (1983).

Bloch, F., "Nuclear Induction," Phys. Rev. **70**, 460 (1946).

Bloch, I., J. Dalibard, and W. Zwerger, "Many-body physics with ultracold gases," Rev. Mod. Phys. **80**, 885 (2008).

Bochinski, J. R., E. R. Hudson, H. J. Lewandowski, G. Meijer, and J. Ye, "Decelerating Neutral Dipolar Molecules," Phys. Rev. Lett. **91**, 243001 (2003).

Boesten, H. M. J. M., A. J. Moerdijk, and B. J. Verhaar, "Dipolar decay in two recent Bose-Einstein condensation experiments," Phys. Rev. A **54**, R29 (1996).

Bohn, J. L., and P. S. Julienne, "Prospects for influencing scattering lengths with far-off-resonant light," Phys. Rev. A **56**, 1486 (1997).

Bohn, J. L., and P. S. Julienne, "Semianalytic theory of laser-assisted resonant cold collisions," Phys. Rev. A **60**, 414 (1999).

Bose, S. N., "Plancks Gesetz und Lichtquantenhypothese," Zeitschrift für Physik **26**, 178 (1924).

Bourdel, T., L. Khaykovich, J. Cubizolles, J. Zhang, F. Chevy, M. Teichmann, L. Tarruell, S. J. J. M. F. Kokkelmans, and C. Salomon, "Experimental study of the BEC-BCS crossover region in lithium 6," Phys. Rev. Lett. **93**, 050401 (2004).

Bradley, C. C., C. A. Sackett, J. J. Tollett, and R. G. Hulet, "Evidence of Bose-Einstein Condensation in an Atomic Gas with Attractive Interactions," Phys. Rev. Lett. **75**, 1687 (1995).

Breit, G., and I. I. Rabi, "Measurement of nuclear spin," Phys. Rev. **38**, 2082 (1931).

Brookhaven, *Brookhaven National Laboratory, National Nuclear Data Center*, http://www.nndc.bnl.gov/nudat2/.

Bussery, B., Y. Achkar, and M. Aubert-Frécon, "Long-range molecular states dissociating to the three or four lowest asymptotes for the ten heteronuclear diatomic alkali molecules," Chem. Phys. **116**, 319 (1987).

Caldas, H., "Cold asymmetrical fermion superfluids," Phys. Rev. A **69**, 063602 (2004).

Camparo, J. C., and R. P. Frueholz, "Parameters of adiabatic rapid passage in the 0-0 hyperne transition of 87Rb," Phys. Rev. A **30**, 803 (1984).

Carlson, J., and S. Reddy, "Asymmetric two-component fermion systems in strong coupling," Phys. Rev. Lett. **95**, 060401 (2005).

Carr, L. D., D. Demille, R. V. Krems, and J. Ye, "Cold and ultracold molecules: science, technology, and applications," New J. Phys. **11**, 055049 (2009).

Casalbuoni, R., and G. Nardulli, "Inhomogeneous superconductivity in condensed matter and QCD," Rev. Mod. Phys. **76**, 263 (2004).

Castin, Y., and R. Dum, "Bose-Einstein condensates in time dependent traps," Phys. Rev. Let. **77**, 5315 (1996).

Chin, C., M. Bartenstein, A. Altmeyer, S. Riedl, S. Jochim, J. H. Denschlag, and R. Grimm, "Observation of the pairing gap in a strongly interacting Fermi gas," Science **305**, 1128 (2004).

Chin, C., V. V. Flambaum, and M. G. Kozlov, "Ultracold molecules: new probes on the variation of fundamental constants," New J. Phys. **11**, 055048 (2009).

Chin, C., R. Grimm, P. Julienne, and E. Tiesinga, "Feshbach Resonances in Ultracold Gases," Rev. Mod. Phys. (in preparation).

Chin, C., and P. S. Julienne, "Radio-frequency transitions on weakly bound ultracold molecules," Phys. Rev. A **71**, 012713 (2005).

Chu, S., in *E. Arimondo, W. Phillips, and F. Strumia (eds.), Proceedings of the International School of Physics "Enrico Fermi", Course CXVIII, p. 239, North-Holland, Amsterdam* (1992).

Chu, S., "Nobel lecture: The manipulation of neutral particles," Rev. Mod. Phys. **70**, 685 (1998).

Chu, S., J. E. Bjorkholm, A. Ashkin, and A. Cable, "Experimental observation of optically trapped atoms," Phys. Rev. Lett. **57**, 314 (1986).

Chu, S., L. Hollberg, J. E. Bjorkholm, A. Cable, and A. Ashkin, "Three-Dimensional Viscous Confinement and Cooling of Atoms by Resonance Radiation Pressure," Phys. Rev. Lett. **55**, 48 (1985).

Chwedeńczuk, J., K. Góral, T. Köhler, and P. S. Julienne, "Molecular Production in Two Component Atomic Fermi Gases," Phys. Rev. Lett. **93**, 260403 (2004).

Clancy, B., L. Luo, and J. E. Thomas, "Observation of Nearly Perfect Irrotational Flow in Normal and Superfluid Strongly Interacting Fermi Gases," Phys. Rev. Lett. **99**, 140401 (2007).

Cohen-Tannoudji, C., B. Diu, and F. Laloë, *Quantum Mechanics, vol. I* (Wiley & Sons, New York) (1977a).

Cohen-Tannoudji, C., B. Diu, and F. Laloë, *Quantum Mechanics, vol. II* (Wiley & Sons, New York) (1977b).

Cohen-Tannoudji, C. N., "Nobel lecture: Manipulating atoms with photons," Rev. Mod. Phys. **70**, 707 (1998).

Cornish, S. L., N. R. Claussen, J. L. Roberts, E. A. Cornell, and C. E. Wieman, "Stable ^{85}Rb Bose-Einstein condensates with widely tunable interactions," Phys. Rev. Lett. **85**, 1795 (2000).

Coursey, J. S., D. J. Schwab, and R. A. Dragoset, *NIST standard reference database 144.* http://physics.nist.gov/PhysRefData/Compositions/.

Courteille, P., R. S. Freeland, D. J. Heinzen, F. A. van Abeelen, and B. J. Verhaar, "Observation of a Feshbach resonance in cold atom scattering," Phys. Rev. Lett. **81**, 69 (1998).

Cubizolles, J., T. Bourdel, S. J. J. M. F. Kokkelmans, G. V. Shlyapnikov, and C. Salomon, "Production of Long-Lived Ultracold Li_2 Molecules from a Fermi Gas," Phys. Rev. Lett. **91**, 240401 (2003).

Dalfovo, F., C. Minniti, S. Stringari, and L. Pitaevskii, "Nonlinear dynamics of a Bose condensed gas," Physics Letters A **227**, 259 (1997).

Dalfovo, F. S., L. P. Pitaevkii, S. Stringari, and S. Giorgini, "Theory of Bose-Einstein condensation in trapped gases," Rev. Mod. Phys. **71**, 463 (1999).

Dalibard, J., in *M. Inguscio, S. Stringari, and C. Wieman (eds.), Proceedings of the International School of Physics Enrico Fermi, Course CXL, p. 321.* IOS Press, Amsterdam (1999).

Dalibard, J., and C. Cohen-Tannoudji, "Dressed-atom approach to atomic motion in laser light: the dipole force revisited," J. Opt. Am. B **2**, 1707 (1985).

Dalibard, J., and C. Cohen-Tannoudji, "Laser cooling below the Doppler limit by polarization gradients: simple theoretical models," J. Opt. Soc. Am. B **6**, 2023 (1989).

Damski, B., L. Santos, E. Tiemann, M. Lewenstein, S. Kotochigova, P. Julienne, and P. Zoller, "Creation of a dipolar superfluid in optical lattices," Phys. Rev. Lett. **90**, 110401 (2003).

Danzl, J. G., E. Haller, M. Gustavsson, M. J. Mark, R. Hart, N. Bouloufa, O. Dulieu, H. Ritsch, and H.-C. Nägerl, "Quantum Gas of Deeply Bound Ground State Molecules," Science **321**, 1062 (2008).

Davis, K. B., M.-O. Mewes, M. R. Andrews, N. J. van Druten, D. M. Kurn, and W. Ketterle, "Bose-Einstein Condensation in a Gas of Sodium Atoms," Phys. Rev. Lett. **75**, 3969 (1995).

De Palo, S., M. Chiofalo, M. Holland, and S. Kokkelmans, "Resonance effects on the crossover of bosonic to fermionic superfluidity," Phys. Lett. A **327**, 490 (2004).

Delannoy, G., S. G. Murdoch, V. Boyer, V. Josse, P. Bouyer, and A. Aspect, "Understanding the production of dual Bose-Einstein condensation with sympathetic cooling," Phys. Rev. A **63**, 051602R (2001).

DeMarco, B., *Quantum Behavior of an Atomic Fermi Gas* (Ph. D. thesis, Graduate School of the University of Colorado) (2001).

DeMarco, B., J. L. Bohn, J. P. Burke, M. Holland, and D. S. Jin, "Measurement of p-Wave Threshold Law Using Evaporatively Cooled Fermionic Atoms," Phys. Rev. Lett. **82**, 4208 (1999a).

DeMarco, B., and D. Jin, "Onset of Fermi degeneracy in a trapped atomic gas," Science **285**, 1703 (1999).

DeMarco, B., S. B. Papp, and D. S. Jin, "Pauli blocking of collisions in a quantum degenerate atomic Fermi gas," Phys. Rev. Lett. **86**, 5409 (2001).

DeMarco, B., H. Rohner, and D. S. Jin, "An enriched ^{40}K source for fermionic atom studies," Rev. Sci. Instrum. **70**, 1967 (1999b).

Desruelle, B., V. Boyer, S. G. Murdoch, G. Delannoy, P. Bouyer, A. Aspect, and M. Lecrivain, "Interrupted evaporative cooling of 87Rb atoms trapped in a high magnetic field," Phys. Rev. A **60**, R1759 (1999).

Dieckmann, K., *Bose-Einstein condensation with high atom number in a deep magnetic trap* (Ph.D. thesis, University of Amsterdam) (2001).

Dieckmann, K., C. A. Stan, S. Gupta, Z. Hadzibabic, C. H. Schunck, and W. Ketterle, "Decay of an Ultracold Fermionic Lithium Gas near a Feshbach Resonance," Phys. Rev. Lett. **89**, 203201 (2002).

D'Incao, J., and B. Esry, "Mass dependence of ultracold three-body collision rates," Phys. Rev. A **73**, 030702(R) (2006).

Drever, R., J. Hall, F. Kowalski, J. Hough, G. Ford, A. Munley, and H. Ward, "Laser phase and frequency stabilization using an optical resonator," Appl. Phys. B **32**, 145 (1983).

Dürr, S., T. Volz, A. Marte, and G. Rempe, "Observation of molecules produced from a Bose-Einstein condensate," Phys. Rev. Lett. **92**, 020406 (2004).

Drullinger, R., D. Wineland, and J. Bergquist, "High-resolution optical spectra of laser cooled ions," Appl. Phys. **22**, 365 (1980).

Duine, R. A., and H. T. C. Stoof, "Dynamics of a Bose-Einstein condensate near a Feshbach resonance," Phys. Rev. A **68**, 013602 (2003).

Eigenwillig, C., *Optimierte Herstellung einer ultrakalten Bose-Fermi-Mischung in der Magnetfalle* (Diplomarbeit, Ludwig-Maximilians University, Munich) (2007).

Einstein, A., "Quantentheorie des idealen einatomigen Gases," Sitzungsber. Preuss. Akad. Wiss. **3**, 18 (1925).

Esslinger, T., I. Bloch, and T. W. Hänsch, "Bose-Einstein condensation in a quadrupole-Ioffe-configuration trap," Phys. Rev. A **58**, R2664 (1998).

Falco, G., and H. Stoof, "Crossover Temperature of Bose-Einstein Condensation in an Atomic Fermi Gas," Phys. Rev. Lett. **92**, 130401 (2004).

Fano, U., "Sullo spettro di assorbimento dei gas nobili presso il limite dello spettro d'arco," Nuovo Cimento **12**, 154 (1935).

Fano, U., "Effects of Configuration Interaction on Intensities and Phase Shifts," Phys. Rev. **124**, 1866 (1961).

Fedichev, P. O., Y. Kagan, G. V. Shlyapnikov, and J. T. M. Walraven, "Influence of Nearly Resonant Light on the Scattering Length in Low-Temperature Atomic Gases," Phys. Rev. Lett. **77**, 2913 (1996a).

Fedichev, P. O., M. W. Reynold, and G. V. Shlyapnikov, "Three-Body Recombination of Ultracold Atoms to a Weakly Bound s Level," Phys. Rev. Lett. **77**, 2921 (1996b).

Feiguin, A. E., and F. Heidrich-Meisner, "Pairing states of a polarized Fermi gas trapped in a one-dimensional optical lattice," Phys. Rev. B **76**, 220508(R) (2007).

Ferlaino, F., C. D'Errico, G. Roati, M. Zaccanti, M. Inguscio, G. Modugno, and A. Simoni, "Erratum: Feshbach spectroscopy of a K-Rb atomic mixture, Phys. Rev. A, 73, 040702, 2006," Phys. Rev. A **74**, 039903 (2006a).

Ferlaino, F., C. D'Errico, G. Roati, M. Zaccanti, M. Inguscio, G. Modugno, and A. Simoni, "Feshbach spectroscopy of a K-Rb atomic mixture," Phys. Rev. A **73**, 040702 (2006b).

Feshbach, H., "Unified theory of nuclear reactions," Ann. Phys. (N. Y.) **5**, 357 (1958).

Feshbach, H., "A unified theory of nuclear reactions II," Ann. Phys. (N. Y.) **19**, 287 (1962).

Flambaum, V. V., C. Gribakin, and C. Harabati, "Analytical calculation of cold-atom scattering," Phys. Rev. A **59**, 1998 (1999).

Fölling, S., F. Gerbier, A. Widera, O. Mandel, T. Gericke, and I. Bloch, "Spatial quantum noise interferometry in expanding ultracold atom clouds," Nature **434**, 481 (2005).

Friebel, S., R. Scheunemann, J. Walz, T. W. Hänsch, and M. Weitz., "Laser cooling in a CO_2-laser optical lattice," Appl. Phys. B **67**, 699 (1998).

Fried, D. G., T. C. Killian, L. Willmann, D. Landhuis, S. C. Moss, D. Kleppner, and T. J. Greytak, "Bose-Einstein condensation of atomic hydrogen," Phys. Rev. Lett. **81**, 3811 (1998).

Fukuhara, T., S. Sugawa, and Y. Takahashi, "Bose-Einstein condensation of an ytterbium isotope," Phys. Rev. A **76**, 051604 (2007a).

Fukuhara, T., Y. Takasu, M. Kumakura, and Y. Takahashi, "Degenerate Fermi gases of ytterbium," Phys. Rev. Lett. **98**, 030401 (2007b).

Fukuhara, T., Y. Takasu, S. S., and Y. Takahashi, "Quantum degenerate Fermi gases of ytterbium atoms," J. Low Temp. Phys. **148**, 441 (2007c).

Fulde, P., and R. A. Ferrell, "Superconductivity in a strong spin-exchange field," Phys. Rev. **135**, A550 (1964).

Fulton, R., A. I. Bishop, M. N. Shneider, and P. F. Barker, "Controlling the motion of cold molecules with deep periodic optical potentials," Nature Physics **2**, 465 (2006).

Gallagher, A., and D. E. Pritchard, "Exoergic Collisions of Cold Na*-Na," Phys. Rev. Lett. **63**, 957 (1989).

Goldman, V. V., and I. F. Silvera, "Atomic hydrogen in an inhomogeneous magnetic field: Density profile and Bose-Einstein condensation," Physical Review B **24**, 2870 (1981).

Goldwin, J., *Quantum Degeneracy and Interactions in the $^{87}Rb - {}^{40}K$ Bose-Fermi Mixture* (Ph.D. thesis, University of Colorado) (2005).

Goldwin, J., S. Inouye, M. L. Olsen, B. Newman, B. D. DePaola, and D. S. Jin, "Measurement of the interaction strength in a Bose-Fermi mixture with ^{87}Rb and ^{40}K," Phys. Rev. A **70**, 021601 (2004).

Góral, K., and L. Santos, "Ground state and elementary excitations of single and binary Bose-Einstein condensates of trapped dipolar gases," Phys. Rev. A **66**, 023613 (2002).

Góral, K., L. Santos, and M. Lewenstein, "Quantum Phases of Dipolar Bosons in Optical Lattices," Phys. Rev. Lett. **88**, 170406 (2002).

Gorkov, L. P., and T. K. Melik-Barkhudarov, "Contribution to the theory of superfluidity in an imperfect Fermi gas," Zh. Eskp. Theor. Fiz. **40**, 1452 [Sov. Phys. JETP **13**, 1018 (1961)] (1961).

Greiner, M., I. Bloch, T. W. Hänsch, and T. Esslinger, "Magnetic transport of trapped cold atoms over a large distance," Phys. Rev. A **63**, 031401 (2001).

Greiner, M., O. Mandel, T. Esslinger, T. W. Häansch, and I. Bloch, "Quantum phase transition from a superfluid to a Mott insulator in a gas of ultracold atoms," Nature **415**, 39 (2002).

Greiner, M., C. A. Regal, and D. S. Jin, "Emergence of a molecular Bose-Einstein condensate from a Fermi gas," Nature **426**, 537 (2003).

Greiner, M., C. A. Regal, J. T. Stewart, and D. S. Jin, "Probing Pair-Correlated Fermionic Atoms through Correlations in Atom Shot Noise," Phys. Rev. Let. **94**, 110401 (2005).

Griesmaier, A., J. Stuhler, T. Koch, M. Fattori, T. Pfau, and S. Giovanazzi, "Comparing contact and dipolar interactions in a Bose-Einstein condensate," Phys. Rev. Lett. **97**, 250402 (2006).

Griesmeier, A., J. Werner, S. Hensler, J. Stuhler, and T. Pfau, "Bose-Einstein condensation of chromium," Phys. Rev. Lett. **94**, 160401 (2005).

Grimm, R., M. Weidemüller, and Y. B. Ovchinnikov, *Optical dipole traps for neutral atoms* (in Advances in Atomic, Molecular and Optical Physics, volume 42, editors B. Bederson and H. Walther, pp. 95–170, Academic Press, San Diego) (2000).

Gupta, S., Z. Hadzibabic, M. W. Zwierlein, C. A. Stan, K. Dieckmann, C. H. Schunck, E. G. M. v. Kempen, B. J. Verhaar, and W. Ketterle, "RF spectroscopy of ultracold fermions," Science **300**, 1723 (2003).

Gurarie, V., and L. Radzihovsky, "Resonantly paired fermionic superfluids," Annals of Physics **322**, 2 (2007).

Hadzibabic, Z., P. Krüger, M. Cheneau, B. Battelier, and J. Dalibard, "Berezinskii-Kosterlitz-Thouless crossover in a trapped atomic gas," Nature **441**, 1118 (2006).

Hadzibabic, Z., C. A. Stan, K. Dieckmann, S. Gupta, M. W. Zwierlein, A. Görlitz, and W. Ketterle, "Two-species mixture of quantum degenerate Bose and Fermi gases," Phys. Rev. Lett. **88**, 160401 (2002).

Hanna, T. M., T. Köhler, and K. Burnett, "Association of molecules using a resonantly modulated magnetic field," Phys. Rev. A **75**, 013606 (2007).

Haussmann, R., W. Rantner, S. Cerrito, and W. Zwerger, "Thermodynamics of the BCS-BEC crossover," Phys. Rev. A **75**, 023610 (2007).

Heiselberg, H., "Fermi systems with long scattering lengths," Phys. Rev. A **63**, 043606 (2001).

Henkel, F., *Fermionisches Kalium in der dreikomponentigen magnetooptischen Falle* (Diplomarbeit, Ludwig-Maximilians University, Munich) (2005).

Herbig, J., T. Kraemer, M. Mark, T. Weber, C. Chin, H.-C. Nägerl, and R. Grimm, "Preparation of a pure molecular quantum gas," Science **301**, 1510 (2003).

Hess, H. F., "Evaporative cooling of magnetically trapped and compressed spin-polarized hydrogen," Phys. Rev. B **34**, 3476 (1986).

Hänsch, T. W., and A. L. Schawlow, "Cooling of gases by laser radiation," Optics Communications **13**, 68 (1975).

Ho, T.-L., "Universal Thermodynamics of Degenerate Quantum Gases in the Unitarity Limit," Phys. Rev. Lett. **92**, 090402 (2004).

Hodby, E., S. T. Thompson, C. A. Regal, M. Greiner, A. C. Wilson, D. S. Jin, E. A. Cornell, and C. E. Wieman, "Production efficiency of ultracold Feshbach molecules in bosonic and fermionic systems," Phys. Rev. Lett. **94**, 120402 (2005).

Hogan, S. D., D. Sprecher, M. Andrist, N. Vanhaecke, and F. Merkt, "Zeeman deceleration of H and D," Phys. Rev. A **76**, 023412 (2007).

Holland, M. J., B. DeMarco, and D. S. Jin, "Evaporative cooling of a two-component degenerate Fermi gas," Phys. Rev. A **61**, 053610 (2000).

Huang, K., *Statistical Mechanics* (John Wiley, New York, 2nd edn.) (1987).

Hudson, E. R., N. B. Gilfoy, S. Kotochigova, J. Sage, and D. DeMille, "Inelastic Collisions of Ultracold Heteronuclear Molecules in an Optical Trap," Phys. Rev. Lett. **100**, 203201 (2008).

Hudson, E. R., H. J. Lewandowski, B. C. Sawyer, and J. Ye, "Cold molecule spectroscopy for constraining the evolution of the fine structure constant," Phys. Rev. Lett. **96**, 143004 (2006a).

Hudson, E. R., C. Ticknor, B. C. Sawyer, C. A. Taatjes, H. J. Lewandowski, J. R. Bochinski, J. L. Bohn, and J. Ye, "Production of cold formaldehyde molecules for study and control of chemical reaction dynamics with hydroxyl radicals," Phys. Rev. A **73**, 063404 (2006b).

Hudson, J. J., B. E. Sauer, M. R. Tarbutt, and E. A. Hinds, "Measurement of the electron electric dipole moment using YbF molecules," Phys. Rev. Lett. **89**, 023003 (2002).

Huse, D. A., and E. D. Siggia, "The density distribution of a weakly interacting Bose gas in an external potential," Journal of Low Temperature Physics **46**, 137 (1982).

Inouye, S., M. Andrews, J. Stenger, H.-J. Miesner, D. M. Stamper-Kurn, and W. Ketterle, "Observation of Feshbach resonances in a Bose-Einstein condensate," Nature **392**, 151 (1998).

Jochim, S., M. Bartenstein, A. Altmeyer, G. Hendl, C. Chin, J. H. Denschlag, and R. Grimm, "Pure gas of optically trapped molecules created from fermionic atoms," Phys. Rev. Lett. **91**, 240402 (2003a).

Jochim, S., M. Bartenstein, A. Altmeyer, G. Hendl, S. Riedl, C. Chin, J. H. Denschlag, and R. Grimm, "Bose-Einstein condensation of molecules," Science **302**, 2101 (2003b).

Jochim, S., M. Bartenstein, G. Hendl, J. H. Denschlag, R. Grimm, A. Mosk, and M. Weidemüller, "Magnetic field control of elastic scattering in a cold gas of fermionic lithium atoms," Phys. Rev. Lett. **89**, 273202 (2002).

Jördens, R., N. Strohmaier, K. Günter, H. Moritz, and T. Esslinger, "A Mott insulator of fermionic atoms in an optical lattice," Nature **455**, 204 (2008).

Julienne, P. S., and J. Vigué, "Cold collisions of ground- and excited-state alkali-metal atoms," Phys. Rev. A **44**, 4464 (1991).

Kagan, Y., E. L. Surkov, and G. V. Shlyapnikov, "Evolution of a Bose-condensed gas under variations of the confining potential," Phys. Rev. Let. **54**, R1753 (1996).

van Kempen, E. G. M., S. J. J. M. F. Kokkelmans, D. J. Heinzen, and B. J. Verhaar, "Interisotopic determination of ultracold rubidium interactions from three high-precision experiments," Phys. Rev. Lett. **88**, 093201 (2002).

Ketterle, W., and N. van Druten, in *B. Bederson and H. Walther (eds.), Advances in Atomic, Molecular, and Optical Physics*, vol. 37, p. 181. Academic Press, San Diego (1996).

Ketterle, W., D. S. Durfee, and D. M. Stamper-Kurn, in *M. Inguscio, S. Stringari, and C. E. Wieman (eds.), Proceedings of the International School of Physics - Enrico Fermi - Course CXL*, p. 67, IOS Press (1999).

Ketterle, W., and M. W. Zwierlein, in *M. Inguscio, W. Ketterle, and C. Salomon (eds.), Ultracold Fermi Gases, Proceedings of the International School of Physics - Enrico Fermi - Course CLXIV, p.95, IOS Press* (2008).

Köhl, M., H. Moritz, T. Stöferle, K. Günter, and T. Esslinger, "Fermionic Atoms in a Three Dimensional Optical Lattice: Observing Fermi Surfaces, Dynamics, and Interactions," Phys. Rev. Lett. **94**, 080403 (2005).

Köhler, T., K. Góral, and P. S. Julienne, "Production of cold molecules via magnetically tunable Feshbach resonances," Rev. Mod. Phys. **78**, 1311 (2006).

Kim, J., B. Friedrich, D. P. Katz, D. Patterson, J. D. Weinstein, R. de Carvalho, and J. M. Doyle, "Buffer-Gas Loading and Magnetic Trapping of Atomic Europium," Phys. Rev. Lett. **78**, 3665 (1997).

Kinast, J., S. L. Hemmer, M. E. Gehm, A. Turlapov, and J. E. Thomas, "Evidence for superfluidity in a resonantly interacting Fermi gas," Phys. Rev. Lett. **92**, 150402 (2004).

Kinoshita, T., T. Wenger, and D. S. Weiss, "Observation of a one-dimensional Tonks-Girardeau gas," Science **305**, 1125 (2004).

Kozlov, M. G., and D. de Mille, "Enhancement of the electric dipole moment of the electron in PbO," Phys. Rev. Lett. **89**, 133001 (2002).

Krems, R. V., "Molecules near absolute zero and external field control of atomic and molecular dynamics," Rev. Phys. Chem. **24**, 99 (2005).

Krems, R. V., "Cold conrolled chemistry," Phys. Chem. Chem. Phys. **10**, 4079 (2008).

Landau, L., and E. Lifschitz, *Quantenmechanik* (vol. III. Akademie Verlag, Berlin) (1985).

Landau, L. D., "On the theory of superfluidity," Phys. Rev. **75**, 884 (1949).

Lang, F., K.Winkler, C. Strauss, R. Grimm, and J. H. Denschlag, "Ultracold Triplet Molecules in the Rovibrational Ground State," Phys. Rev. Lett. **101**, 133005 (2008).

Larkin, A. I., and Y. N. Ovchinnikov Sov. Phys. JETP **20**, 762 (1965).

Larson, D. J., J. C. Bergquist, J. J. Bollinger, W. M. Itano, and D. J. Wineland, "Sympathetic Cooling of Trapped Ions: A Laser-Cooled Two-Species Nonneutral Ion Plasma," Phys. Rev. Lett. **57**, 70 (1986).

Leggett, A. J., "Bose-Einstein condensation in the alkali gases: Some fundamental concepts," Rev. Mod. Phys. **73**, 307 (2001).

Li, Z., S. Singh, T. V. Tscherbul, and K. W. Madison, "Feshbach resonances in ultracold ^{85}Rb $-^{87}$Rb and ^{6}Li $-^{87}$Rb mixtures," Phys. Rev. A **78**, 022710 (2008).

Liu, W. V., and F. Wilczek, "Interior gap superfluidity," Phys. Rev. Lett. **90**, 047002 (2003).

Loftus, T., C. A. Regal, C. Ticknor, J. L. Bohn, and D. S. Jin, "Resonant control of elastic collisions in an optical trapped Fermi gas of atoms," Phys. Rev. Lett. **88**, 173201 (2002).

London, F., "On the Bose-Einstein Condensation," Phys. Rev. **54**, 947 (1938a).

London, F., "The λ-phenomenon of liquid helium and the Bose-Einstein degeneracy," Nature **141**, 643 (1938b).

Luiten, O. J., M. W. Reynolds, and J. T. M. Walraven, "Kinetic theory of the evaporative cooling of a trapped gas," Phys. Rev. A **53**, 381 (1996).

Majorana, E., "Atomi orientati in campo magnetico variabile," Nuovo Cimento **9**, 43 (1932).

Marte, A., T. Volz, J. Schuster, S. Dürr, G. Rempe, E. G. M. van Kempen, and B. J. Varhaar, "Feshbach resonances in rubidium 87: Precision measurement and analysis," Phys. Rev. Lett. **89**, 283202 (2002).

Martin, A. G., K. Helmerson, V. S. Bagnato, G. P. Lafyatis, and D. E. Pritchard, "rf Spectroscopy of Trapped Neutral Atoms," Phys. Rev. Lett. **61**, 2431 (1988).

Masuhara, N., J. M. Doyle, J. C. Sandberg, D. Kleppner, T. J. Greytak, H. F. Hess, and G. P. Kochanski, "Evaporative Cooling of Spin-Polarized Atomic Hydrogen," Phys. Rev. Lett. **61**, 935 (1988).

McAlexander, W. I., E. R. I. Abraham, and R. G. Hulet, "Radiative lifetime of the 2 P state of lithium," Phys. Rev. A **54**, R5 (1996).

McNamara, J. M., T. Jeltes, A. S. Tychkov, W. Hogervorst, and W. Vassen, "Degenerate Bose-Fermi mixture of metastable atoms," Phys. Rev. Lett. **97**, 080404 (2006).

Metcalf, H. J., and P. van der Straaten, *Laser Cooling and Trapping* (Springer, New York) (2002).

Micheli, A., G. K. Brennen, and P. Zoller, "A toolbox for lattice-spin models with polar molecules," Nature Physics **2**, 341 (2006).

Mies, F. H., E. Tiesinga, and P. S. Julienne, "Manipulation of Feshbach resonances in ultracold atomic collisions using time-dependent magnetic fields," Phys. Rev. A **61**, 022721 (2000).

Mies, F. H., C. J. Williams, P. S. Julienne, and M. Krauss, "Estimating bounds on collisional relaxation rates of spin-polarized ^{87}Rb atoms at ultracold temperatures," J. Res. Natl. Inst. Stand. Technol. **101**, 521 (1996).

Migdall, A. L., J. V. Prodan, W. D. Phillips, T. H. Bergemann, and H. J. Metcalf, "First observation of magnetically trapped neutral atoms," Phys. Rev. Lett. **54**, 2596 (1985).

de Mille, D., "Quantum computation with trapped polar molecules," Phys. Rev. Lett. **88**, 067901 (2002).

Modugno, G., F. Ferlaino, R. Heidemann, G. Roati, and M. Inguscio, "Production of a Fermi gas of atoms in an optical lattice," Phys. Rev. A **68**, 011601(R) (2003).

Modugno, G., G. Ferrari, G. Roati, R. J. Brecha, A. Simoni, and M. Inguscio, "Bose-Einstein condensation of potassium atoms by sympathetic cooling," Science **294**, 1320 (2001).

Moerdijk, A. J., and B. J. Verhaar, "Collisional two- and three-body decay rates of dilute quantum gases at ultralow temperatures," Phys. Rev. A **53**, R19 (1996).

Moerdijk, A. J., B. J. Verhaar, and A. Axelsson, "Resonances in ultracold collisions of ^{6}Li, ^{7}Li, and ^{23}Na," Phys. Rev. A **51**, 4852 (1995).

Mosk, A., S. Kraft, M. Mudrich, K. Singer, W. Wohlleben, R. Grimm, and M. Weidemüller, "Mixture of ultracold lithium and cesium atoms in an optical dipole trap," Appl. Phys. B **73**, 791 (2001).

Movre, M., and R. Beuc, "van der Waals interaction in excited alkali-metal dimers," Phys. Rev. A **31**, 2957 (1984).

Myatt, C. J., E. A. Burt, R. W. Ghrist, E. A. Cornell, and C. E. Wieman, "Production of Two Overlapping Bose-Einstein Condensates by Sympathetic Cooling," Phys. Rev. Lett. **78**, 586 (1997).

Narevicius, E., A. Libson, C. G. Parthey, I. Chavez, J. Narevicius, U. Even, and M. G. Raizen, "Stopping Supersonic Beams with a Series of Pulsed Electromagnetic Coils: An Atomic Coilgun," Phys. Rev. Lett. **100**, 093003 (2008).

Ni, K.-K., S. Ospelkaus, M. H. G. de Miranda, A. Pe'er, B. Neyenhuis, J. J. Zirbel, S. Kotochigova, P. S. Julienne, D. S. Jin, and J. Ye, "A High Phase-Space-Density Gas of Polar Molecules," Science **322**, 231 (2008).

Nozières, P., and S. Schmitt-Rink, "Bose condensation in an attractive fermion gas: From weak to strong coupling superconductivity," J. Low Temp. Phys. **59**, 195 (1985).

O'Hara, K. M., *Optical Trapping and Evaporative Cooling of Fermionic Atoms* (Ph.D. thesis, Duke University) (2000).

O'Hara, K. M., S. L. Hemmer, M. E. Gehm, S. R. Granade, and J. E. Thomas, "Observation of a Strongly Interacting Degenerate Fermi Gas of Atoms," Science **298**, 2179 (2002a).

O'Hara, K. M., S. L. Hemmer, S. R. Granade, M. E. Gehm, J. E. Thomas, V. Venturi, E. Tiesinga, and C. J. Williams, "Measurement of the zero crossing in a Feshbach resonance of fermionic ^{6}Li," Phys. Rev. A **66**, 041401 (2002b).

Ohashi, Y., and A. Griffin, "Superfluid transition temperature in a trapped gas of Fermi atoms with a Feshbach resonance," Phys. Rev. A **67**, 033603 (2003).

Orso, G., "Attractive Fermi Gases with Unequal Spin Populations in Highly Elongated Traps," Phys. Rev. Lett. **98**, 070402 (2007).

Ospelkaus, C., S. Ospelkaus, L. Humbert, P. Ernst, K. Sengstock, and K. Bongs, "Ultracold heteronuclear molecules in a 3D optical lattice," Phys. Rev. Lett. **97**, 120402 (2006a).

Ospelkaus, C., S. Ospelkaus, K. Sengstock, and K. Bongs, "Interaction-driven dynamics of ^{40}K$-^{87}$Rb Fermion-Boson gas mixtures in the large-particle-number limit," Phys. Rev. Lett. **96**, 020401 (2006b).

Bibliography

Ospelkaus, S., C. Ospelkaus, L. Humbert, K. Sengstock, and K. Bongs, "Tuning of Heteronuclear Interactions in a Degenerate Fermi-Bose Mixture," Phys. Rev. Lett. **97**, 120403 (2006c).

Ospelkaus, S., A. Pe'er, K.-K. Ni, J. J. Zirbel, B. Neyenhuis, S. Kotochigova, P. S. Julienne, J. Ye, and D. S. Jin, "Efficient state transfer in an ultracold dense gas of heteronuclear molecules," Nature Physics **4**, 622 (2008).

Papp, S. B., and C. E. Wieman, "Observation of heteronuclear Feshbach molecules from a ^{85}Rb $-^{87}$Rb gas," Phys. Rev. Lett. **97**, 180404 (2006).

Paredes, B., A. Widera, V. Murg, O. Mandel, S. Fölling, I. Cirac, G. V. Shlyapnikov, T. W. Hänsch, and I. Bloch, "Tonks-Girardeau gas of ultracold atoms in an optical lattice," Nature **429**, 277 (2004).

Partridge, G. B., W. Li, R. I. Kamar, Y. Liao, and R. G. Hulet, "Pairing and phase separation in a polarized Fermi gas," Science **311**, 503 (2006).

Pe'er, A., E. A. Shapiro, M. C. Stowe, M. Shapiro, and J. Ye, "Precise control of molecular dynamics with a femtosecond frequency comb - a weak field route to strong field coherent control," Phys. Rev. Lett. **98**, 113004 (2007).

Pethick, C., and H. Smith, *Bose-Einstein Condensation in Dilute Gases* (Cambridge University Press, Cambridge, U.K.) (2002).

Petrich, W., M. H. Anderson, J. R. Ensher, and E. A. Cornell, "Bahavior of atoms in a compressed magneto-optical trap," J. Opt. Soc. Am. B **11**, 1332 (1994).

Petrich, W., M. H. Anderson, J. R. Ensher, and E. A. Cornell, "Stable, Tightly Confining Magnetic Trap for Evaporative Cooling of Neutral Atoms," Phys. Rev. Lett. **74**, 3352 (1995).

Petrov, D. S., "Three-Boson Problem near a Narrow Feshbach Resonance," Phys. Rev. Lett. **93**, 143201 (2004).

Petrov, D. S., G. E. Astrakharchik, D. J. Papoular, C. Salomon, and G. V. Shlyapnikov, "Crystalline phase of strongly interacting Fermi mixtures," Phys. Rev. Lett. **99**, 130407 (2007).

Petrov, D. S., C. Salomon, and G. V. Shlyapnikov, "Weakly bound dimers of fermionic atoms," Phys. Rev. Lett. **93**, 090404 (2004).

Petrov, D. S., C. Salomon, and G. V. Shlyapnikov, "Diatomic molecules in ultracold Fermi gases - novel composite bosons," J. Phys. B **38**, S645 (2005).

Phillips, W. D., "Nobel lecture: Laser cooling and trapping of neutral atoms," Rev. Mod. Phys. **70**, 721 (1998).

Phillips, W. D., and H. Metcalf, "Laser deceleration of an atomic beam," Phys. Rev. Lett. **48**, 596 (1982).

Pitaevskii, L., and S. Stringari, *Bose-Einstein Condensation* (Oxford University Press, Oxford) (2003).

Presilla, C., and R. Onofrio, "Cooling Dynamics of Ultracold Two-Species Fermi-Bose Mixtures," Phys. Rev. Lett. **90**, 030404 (2003).

Pritchard, D., K. Helmerson, and A. Martin, in *edited by S. Haroche, J.C. Gay, and G. Grynberg (World Scientic, Singapore), p. 1* (1989).

Pritchard, D. E., "Cooling Neutral Atoms in a Magnetic Trap for Precision Spectroscopy," Phys. Rev. Lett. **51**, 1336 (1983).

Raab, E. L., M. Prantiss, A. Cable, S. Chu, and D. E. Pritchard, "Trapping of neutral sodium atoms with radiation pressure," Phys. Rev. Lett. **59**, 2631 (1987).

Rapp, A., G. Zaránd, C. Honerkamp, and W. Hofstetter, "Color Superfluidity and "Baryon" Formation in Utracold Fermions," Phys. Rev. Lett. **98**, 160405 (2007).

Regal, C. A., M. Greiner, and D. S. Jin, "Lifetime of Molecule-Atom Mixtures near a Feshbach Resonance in ^{40}K," Phys. Rev. Lett. **92**, 083201 (2004a).

Regal, C. A., M. Greiner, and D. S. Jin, "Observation of resonance condensation of fermionic atom pairs," Phys. Rev. Lett. **92**, 040403 (2004b).

Regal, C. A., C. Ticknor, J. L. Bohn, and D. S. Jin, "Creation of ultracold molecules from a Fermi gas of atoms," Nature **424**, 47 (2003).

Ricci, L., M. Weidemüller, T. Esslinger, A. Hemmerich, C. Zimmermann, V. Vuletic, W. König, and T. W. Hänsch, "A compact grating-stabilized diode laser system for atomic physics," Optics Comm. **117**, 541 (1995).

Roati, G., W. Jastrzebski, A. Simoni, G. Modugno, and M. Inguscio, "Optical trapping of cold fermionic potassium for collisional studies," Phys. Rev. A **63**, 052709 (2001).

Roati, G., F. Riboli, G. Modugno, and M. Inguscio, "Fermi-Bose quantum degenerate 40K-87Rb mixture with attractive interaction," Phys. Rev. Lett. **89**, 150403 (2002).

Robert, A., O. Sirjean, A. Browaeys, J. Poupard, S. Nowak, D. Boiron, C. I. Westbrook, and A. Aspect, "A Bose-Einstein condensate of metastable atoms," Science **292**, 461 (2001).

Roberts, J. L., N. R. Claussen, J. P. Burke, C. H. Greene, E. A. Cornell, and C. E. Wieman, "Resonant magnetic field control of elastic scattering in cold ^{85}Rb," Phys. Rev. Lett. **81**, 5109 (1998).

Roberts, J. L., N. R. Claussen, S. L. Cornish, and C. E. Wieman, "Magnetic field dependence of ultracold inelastic collisions near a Feshbach resonance," Phys. Rev. Lett. **85**, 728 (2000).

Rom, T., T. Best, D. van Oosten, U. Schneider, S. Folling, B. Paredes, and I. Bloch, "Free fermion antibunching in a degenerate atomic Fermi gas released from an optical lattice," Nature **444**, 733 (2006).

Rosa, M. D. D., "Laser-cooling molecules," Eur. Phys. J D **31**, 395 (2004).

Rubbmark, J. R., M. M. Kash, M. G. Littman, and D. Kleppner, "Dynamical effects at avoided level crossing: A study of the Landau-Zener effect using rydberg atoms," Physical Review A **23**, 3107 (1981).

Safronova, M. S., B. Arora, and C. W. Clark, "Frequency-dependent polarizabilities of alkali-metal atoms from ultraviolet through infrared spectral regions," Phys. Rev. A **73**, 022505 (2006).

Sage, J. M., S. Sainis, T. Bergeman, and D. DeMille, "Optical Production of Ultracold Polar Molecules," Phys. Rev. Lett. **94**, 203001 (2005).

Sakurai, J., *Modern quantum mechanics* (Addison-Wesley, New York, 2nd edn.) (1994).

Santos, F. P. D., J. Léonard, J. Wang, C. J. Barrelet, F. Perales, E. Rasel, C. S. Unnikrishnan, M. Leduc, and C. Cohen-Tannoudji, "Bose-Einstein condensation of metastable helium," Phys. Rev. Lett. **86**, 3459 (2001).

Santos, L., G. V. Shlyapnikov, P. Zoller, and M. Lewenstein, "Bose-Einstein condensation in trapped dipolar gases," Phys. Rev. Lett. **85**, 1791 (2000).

Santos, L., G. V. Shlyapnikov, P. Zoller, and M. Lewenstein, "Erratum: Bose-Einstein condensation in trapped dipolar gases [Phys. Rev. Lett. 85, 1791 (2000)]," Phys. Rev. Lett. **88**, 139904 (2002).

Sarma, G., "On the influence of a uniform exchange field acting on the spins of the conduction electrons in a superconductor," J. Phys. Chem. Solids **24**, 1029 (1963).

Scherf, W., O. Khait, H. Jäger, and L. Windholz, "Re-measurement of the transition frequencies, fine structure splitting and isotopic shift of the resonance lines of lithium, sodium and potassium," Z. Phys. D **36**, 31 (1996).

Schneider, U., L. Hackermüller, S. Will, T. Best, I. Bloch, T. A. Costi, R. W. Helmes, D. Rasch, and A. Rosch, "Metallic and Insulating Phases of Repulsively Interacting Fermions in a 3D Optical Lattice," Science **322**, 1520 (2008).

Schünemann, U., H. Engler, R. Grimm, M. Weidemüller, and M. Zielonkowski, "Simple scheme for tunable frequency offset locking of two lasers," Rev. Sci. Instrum. **70**, 242 (1999).

Schreck, F., L. Khaykovich, K. L. Corwin, G. Ferrari, T. Bourdel, J. Cubizolles, and C. Salomon, "Quasipure Bose-Einstein condensate immersed in a Fermi sea," Phys. Rev. Lett. **87**, 080403 (2001).

Schunck, C. H., Y. Shin, A. Schirotzek, M. W. Zwierlein, and W. Ketterle, "Pairing Without Superfluidity: The Ground State of an Imbalanced Fermi Mixture," Science **316**, 867 (2007).

Schunck, C. H., M. W. Zwierlein, C. A. Stan, S. M. F. Raupach, W. Ketterle, A. Simoni, E. Tiesinga, C. J. Williams, and P. S. Julienne, "Feshbach resonances in fermionic ^{6}Li," Phys. Rev. A **71**, 045601 (2005).

Schweikhard, V., S. Tung, and E. A. Cornell, "Vortex Proliferation in the Berezinskii-Kosterlitz-Thouless Regime on a Two-Dimensional Lattice of Bose-Einstein Condensates," Phys. Rev. Lett. **99**, 030401 (2007).

Bibliography

Shapiro, E. A., A. Pe'er, J. Ye, and M. Shapiro, "Piecewise adiabatic population transfer in a molecule via a wave packet," Phys. Rev. Lett. **101**, 023601 (2008).

Sheehy, D. E., and L. Radzihovsky, "BEC-BCS crossover, phase transitions and phase separation in polarized resonantly-paired superfluids," Annals of Physics **322**, 1790 (2007).

Silber, C., S. Günther, C. Marzok, B. Deh, P. W. Courteille, and C. Zimmermann, "Quantum-degenerate mixture of fermionic lithium and bosonic rubidium gases," Phys. Rev. Lett. **95**, 170408 (2005).

Simonucci, S., P. Pieri, and G. C. Strinati, "Broad vs. narrow Fano-Feshbach resonances in the BCS-BEC crossover with trapped Fermi atoms," Europhys. Lett. **69**, 713 (2005).

Stan, C. A., M. W. Zwierlein, C. H. Schunck, S. M. F. Raupach, and W. Ketterle, "Observation of Feshbach Resonances between Two Different Atomic Species," Phys. Rev. Lett. **93**, 143001 (2004).

Stenger, J., S. Inouye, M. R. Andrews, H.-J. Miesner, D. M. Stamper-Kurn, and W. Ketterle, "Strongly enhanced inelastic collisions in a Bose-Einstein condensate near Feshbach resonances," Phys. Rev. Lett. **82**, 2422 (1999).

Sterr, U., "Bose-Einstein Condensation of Calcium," private communication (2009).

Stoof, H. T. C., J. M. V. A. Koelman, and B. J. Verhaar, "Spin-exchange and dipole relaxation rates in atomic hydrogen: Rigorous and simplified calculations," Phys. Rev. B **38**, 4688 (1988).

Strecker, K. E., G. B. Partridge, and R. Hulet, "Conversion of an Atomic Fermi Gas to a Long-Lived Molecular Bose Gas," Phys. Rev. Lett. **91**, 080406 (2003).

Stuhl, B. K., B. C. Sawyer, D. Wang, and J. Ye, "Magneto-optical Trap for Polar Molecules," Phys. Rev. Lett. **101**, 243002 (2008).

Taglieber, M., *Quantum Degeneracy in an Atomic Fermi-Fermi-Bose Mixture* (Dissertation, Ludwig-Maximilians University, Munich) (2008).

Taglieber, M., A.-C. Voigt, T. Aoki, T. W. Hänsch, and K. Dieckmann, "Quantum Degenerate Two-Species Fermi-Fermi Mixture Coexisting with a Bose-Einstein Condensate," Phys. Rev. Lett. **100**, 010401 (2008).

Taglieber, M., A.-C. Voigt, F. Henkel, S. Fray, T. W. Hänsch, and K. Dieckmann, "Simultaneous magneto-optical trapping of three atomic species," Phys. Rev. A **73**, 011402 (2006).

Takasu, Y., K. Maki, K. Komori, T. Takano, K. Honda, M. Kumakura, T. Yabuzaki, and Y. Takahashi, "Spin-singlet Bose-Einstein condensation of two-electron atoms," Phys. Rev. Lett. **91**, 040404 (2003).

Taylor, J., *Scattering Theory* (Wiley, New York) (1972).

Thalhammer, G., K. Winkler, F. Lang, S. Schmid, R. Grimm, and J. H. Denschlag, "Long-lived Feshbach molecules in a three-dimensional optical lattice," Phys. Rev. Lett. **96**, 050402 (2006).

Bibliography

Thompson, S. T., E. Hodby, and C. E. Wieman, "Ultracold Molecule Production via a Resonant Oscillating Magnetic Field," Phys. Rev. Lett. **95**, 190404 (2005).

Thorsheim, H. R., J. Weiner, and P. S. Julienne, "Laser-induced photoassociation of ultracold sodium atoms," Phys. Rev. Lett. **58**, 2420 (1987).

Tiemann, E., H. Knockel, P. Kowalczyk, W. Jastrzebski, A. Pashov, H. Salami, and A. J. Ross, "Coupled system $a^3\Sigma^+$ and $X^1\Sigma^+$ of KLi: Feshbach resonances and corrections to the Born-Oppenheimer approximation," Phys. Rev. A **79**, 042716 (2009).

Tiesinga, E., B. J. Verhaar, and H. T. C. Stoof, "Threshold and resonance phenomena in ultracold ground-state collisions," Phys. Rev. A **47**, 4114 (1993).

Timmermans, E., and R. Côté, "Superfluidity in Sympathetic Cooling with Atomic Bose-Einstein Condensates," Phys. Rev. Lett. **80**, 3419 (1998).

Timmermans, E., P. Tommasini, M. Hussein, and A. Kerman, "Feshbach resonances in atomic Bose-Einstein condensates," Phys. Rep. **315**, 199 (1999).

Truscott, A. G., K. E. Strecker, W. I. McAlexander, G. B. Partridge, and R. G. Hulet, "Observation of Fermi pressure in a gas of trapped atoms," Science **291**, 2570 (2001).

Ungar, P. J., D. S. Weiss, E. Riis, and S. Chu, "Optical molasses and multilevel atoms: theory," J. Opt. Soc. Am. B **6**, 2058 (1989).

Vadla, C., C. J. Lorenzen, and K. Niemax, "Repulsive van der Waals and Dipole-Quadrupole Interaction in the Excited LiCs Molecule," Phys. Rev. Lett. **51**, 988 (1983).

Valkering, P., *Optimization of evaporative cooling of Rubidium atoms in a magnetic trap* (Master's thesis, FOM Institute for Atomic and Molecular Physics, Amsterdam, Unpublished) (1999).

Voigt, A.-C., *Simultanes Fangen von drei Atomsorten in einer magnetooptischen Falle* (Diplomarbeit, Technical University of Munich, Munich, Unpublished) (2004).

Voigt, A.-C., M. Taglieber, L. Costa, T. Aoki, W. Wieser, T. W. Hänsch, and K. Dieckmann, "Ultrocold Heteronuclear Fermi-Fermi Molecules," Phys. Rev. Lett. **102**, 020405 (2009).

Volz, T., N. Syassen, D. M. Bauer, E. Hansis, S. Dürr, and G. Rempe, "Preparation of a quantum state with one molecule at each site of an optical lattice," Nature Physics **2**, 692 (2006).

Volz, U., and H. Schmoranzer, "Precision lifetime measurement on alkali atoms and on helium by beam-gas-laser spectroscopy," Phys. Scr. **T65**, 48 (1996).

Vuletić, V., A. J. Kerman, C. Chin, and S. Chu, "Observation of low-field Feshbach resonances in collisions of cesium atoms," Phys. Rev. Lett. **82**, 1406 (1999).

Walker, T., and P. Feng (1994).

Walker, T., D. Sesko, and C. Wieman Phys. Rev. Lett. **64**, 408 (1990).

Walraven, J., in *G. L. Oppo, S. M. Barnett, E. Riis, and M. Wilkinson (eds.), Quantum Dynamics of Simple Systems, CRC Press, London* (1996).

Walraven, J., *Elements of Quantum Gases: Thermodynamic and Collisional Properties of Trapped Atomic Gases* (University of Amsterdam) (2009).

Wang, B., H.-D. Chen, and S. Das Sarma, "Quantum phase diagram of fermion mixtures with population imbalance in one-dimensional optical lattices," arXiv:quant-ph 0901.4896v1 (2009).

Wang, H., and W. C. Stwalley, "Ultracold photoassociative spectroscopy of heteronuclear alkali-metal diatomic molecules," J. Chem. Phys. **108**, 5767 (1998).

Wasik, G., W. Gawlik, J. Zachorowski, and W. Zawadzki, "Laser frequency stabilization by Doppler-free magnetic dichroism," Appl. Phys. B **75**, 613 (2002).

Weber, C., G. Barontini, J. Catani, G. Thalhammer, M. Inguscio, and F. Minardi, "Association of ultracold double-species bosonic molecules," Phys. Rev. A **78**, 061601(R) (2008).

Weber, T., J. Herbig, M. Mark, H.-C. Nägerl, and R. Grimm, "Bose-Einstein condensation of cesium," Science **299**, 232 (2002).

Weber, T., J. Herbig, M. Mark, H.-C. Nägerl, and R. Grimm, "Three-Body Recombination at Large Scattering Lengths in an Ultracold Atomic Gas," Phys. Rev. Lett. **91**, 123201 (2003).

Weiner, J. (1995).

Weiner, J., V. S. Bagnato, S. Zilio, and P. S. Julienne, "Experiments and theory in cold and ultracold collisions," Rev. Mod. Phys. **71**, 1 (1999).

Weinstein, J. D., R. deCarvalho, T. Guillet, B. Friedrich, and J. Doyle, "Magnetic trapping of calcium monohydride molecules at millikelvin temperatures," Nature **395**, 148 (1998).

Wieman, C. E., and L. Hollberg, "Using diode lasers for atomic physics," Rev. Sci. Instrum. **62**, 1 (1991).

Wieser, W., *An Optical Dipole Trap for Ultracold Bosons and Fermions* (Diplomarbeit, Ludwig-Maximilians University, Munich, Unpublished) (2006).

Wilczek, D., "Quantum chromodynamics: Lifestyles of the small and simple," Nature Physics **3**, 375 (2007).

Wille, E., F. M. Spiegelhalder, G. Kerner, D. Naik, A. Trenkwalder, G. Hendl, F. Schreck, R. Grimm, T. G. Tiecke, J. T. M. Walraven, S. J. J. M. F. Kokkelmans, E. Tiesinga, *et al.*, "Exploring an ultracold Fermi-Fermi mixture: Interspecies Feshbach resonances and scattering properties of ^{6}Li and ^{40}K," Phys. Rev. Lett. **100**, 053201 (2008).

Windholz, L., and M. Musso, "Zeeman- and Paschen-Back-effect of the hyperfine structure of the sodium D_2-line," Z. Phys. D - Atoms, Molecules and Clusters **8**, 239 (1988).

Wineland, D. J., R. E. Drullinger, and F. L. Walls, "Radiation-pressure cooling of bound resonant absorbers," Phys. Rev. Lett. **40**, 1639 (1978).

Wing, W. H., "On neutral particle trapping in quasistatic electromagnetic fields," Prog. Quant. Electr. **8**, 181 (1984).

Wright, M. J., S. Riedl, A. Altmeyer, C. Kohstall, E. R. Sánchez Guajardo, J. Hecker Denschlag, and R. Grimm, "Finite-Temperature Collective Dynamics of a Fermi Gas in the BEC-BCS Crossover," Phys. Rev. Lett. **99**, 150403 (2007).

Xu, K., T. Mukaiyama, J. R. Abo-Shaeer, J. K. Chin, D. E. Miller, and W. Ketterle, "Formation of quantum-degenerate sodium molecules," Phys. Rev. Lett. **91**, 210402 (2003).

Ye, J., S. Schwartz, P. Jungner, and J. L. Hall, "Hyperfine structure and absolute frequency of the ^{87}Rb 5 $P_{3/2}$ state," Opt. Lett. **21**, 1280 (1996).

Zener, C., "Non-adiabatic crossing of energy levels," Proceedings of the Royal Society of London Series A **137**, 696 (1932).

Zirbel, J. J., K.-K. Ni, S. Ospelkaus, J. P. D'Incao, C. E. Wieman, J. Ye, and D. S. Jin, "Heteronuclear molecules in an optical dipole trap," Phys. Rev. Lett. **100**, 143201 (2008a).

Zirbel, J. J., K.-K. Ni, S. Ospelkaus, T. L. Nicholson, M. L. Olsen, C. E. Wieman, J. Ye, D. S. Jin, and P. S. Julienne, "Heteronuclear molecules in an optical dipole trap," Phys. Rev. A **78**, 013416 (2008b).

Zwierlein, M. W., J. R. Abo-Shaeer, A. Schirotzek, C. H. Schunck, and W. Ketterle, "Vortices and superfluidity in a strongly interacting Fermi gas," Nature **435**, 1047 (2005).

Zwierlein, M. W., A. Schirotzek, C. H. Schunck, and W. Ketterle, "Fermionic superfluidity with imbalanced spin populations," Science **311**, 492 (2006).

Zwierlein, M. W., C. A. Stan, C. H. Schunck, S. M. F. Raupach, S. Gupta, Z. Hadzibabic, and W. Ketterle, "Observation of Bose-Einstein condensation of molecules," Phys. Rev. Lett. **91**, 250401 (2003).

Zwierlein, M. W., C. A. Stan, C. H. Schunck, S. M. F. Raupach, A. J. Kerman, and W. Ketterle, "Condensation of pairs of fermionic atoms near a Feshbach resonance," Phys. Rev. Lett. **92**, 120403 (2004).

Bibliography

Danksagung

An erster Stelle danke ich ganz besonders meinem Doktorvater Herrn Professor Dr. Theodor W. Hänsch. Es war mir eine persönliche Freude und Ehre Mitglied seiner Gruppe sein zu dürfen und in seinem stets motivierenden und fruchtenden Umfeld mitzuarbeiten, was mich immer wieder vorangetrieben hat.

Herrn Prof. Dr. Wilhelm Zwerger danke ich ausdrücklich für seine Bereitschaft, die vorliegende Arbeit zu begutachten.

Ich danke meinem Gruppenleiter Herrn Dr. Kai Dieckmann, der mich in die faszinierende Welt der Quantengase eingeführt hat und durch den ich das Experimentieren mit den modernsten und komplexesten Apparaten von der Pieke auf gelernt habe. Sein stets offenes Ohr für meine "wildesten" Ideen waren sehr hilfreich.

Matthias Taglieber, der mit mir das vorliegende Experiment aufgebaut hat, danke ich für die ausgezeichnete Zusammenarbeit. Mit ihm habe ich gemeinsam den Großteil der langen Labornächte verbracht und alle Hochs und Tiefs des Experiments durchlaufen. Seine besondere Eigenschaft in dieser Zeit war, dass er immer die schwierigsten Zusammenhänge bildlich auf den Punkt bringen konnte.

Louis Costa danke ich für die prima Zusammenarbeit zum Schluss meiner Arbeit. Er hat sich außergewöhnlich schnell in das sehr komplexe Experiment einarbeiten können und war mir immer ein bereichernder Diskussionspartner.

Für die besonders netten Stunden mit ihm und seinen Geschichten aus Fernost bin ich unserem japanischen Post-Doc Takatoshi Aoki dankbar. Die exotischen Mitbringsel aus seiner Heimat haben mich und uns alle im Team immer sehr fasziniert.

Darüber hinaus möchte ich mich bei unseren Diplomanden Wolfgang Wieser, Christoph Eigenwillig, Florian Henkel und unserem ehemaligen Post-Doc Sebastian Fray bebedanken, die in besonderer Weise zum Gelingen dieser Arbeit mit beigetragen haben.

Außerdem gebührt Dank unseren Werkstudenten Benedikt Breitenfeld, Norbert Ortegel, Caroline Hahn, Peter Hilz, Johannes Hofer, Johannes Thüringen, Matthias Mang und Gabriel Bismut. Ohne sie wäre die viele Arbeit im Labor nie zu bewältigen gewesen.

Zu besonderen Dank bin ich auch Herrn Toni Scheich verpflichtet, der mir auch die schwierigsten elektronischen Schaltungen erklären konnte und mir immer mit Rat und Tat zur Seite

DANKSAGUNG

stand.

Der Techniker-Mannschaft am MPQ um Wolfgang Simon, Charlie Linner und Helmut Brückner danke ich für die erstklassige Unterstützung beim Design der Apparatur und bei vielen technischen Fragen. Auch möchte ich der gesamten Belegschaft der LMU- als auch der MPQ-Werkstatt für deren ausgezeichnete Arbeit danken.

Frau Gabriele Gschwendtner hat für ihr organisatorisches Geschick besonderen Dank verdient. In beeindruckender Weise hat sie während meiner Zeit als Doktorand die Abteilung gemanagt. Frau Nicole Schmidt möchte ich für die technische Unterstützung im Verlauf der Erstellung meiner Doktorarbeit danken.

Überdies danke ich unseren anderen Arbeitsgruppen: der Arbeitsgruppe von Professor Dr. Weinfurter, der Mikrofallen Gruppe von Dr. Treutlein und der gesamten Arbeitsgruppe am MPQ. Sie standen immer für neue Diskussionen bereit und konnten mir in Notsituationen mit optischen Komponenten stets aushelfen.

Besonders möchte ich mich von ganzem Herzen bei meinen Eltern und meiner Schwester bedanken, die mich auf meinem bisherigen Lebens- und Ausbildungsweg stets tatkräftig mit viel Geduld, Zuspruch und Vertrauen unterstützt haben.

Die VDM Verlagsservicegesellschaft sucht für wissenschaftliche Verlage abgeschlossene und herausragende

Dissertationen, Habilitationen, Diplomarbeiten, Master Theses, Magisterarbeiten usw.

für die kostenlose Publikation als Fachbuch.

Sie verfügen über eine Arbeit, die hohen inhaltlichen und formalen Ansprüchen genügt, und haben Interesse an einer honorarvergüteten Publikation?

Dann senden Sie bitte erste Informationen über sich und Ihre Arbeit per Email an *info@vdm-vsg.de*.

Sie erhalten kurzfristig unser Feedback!

VDM Verlagsservicegesellschaft mbH
Dudweiler Landstr. 99
D - 66123 Saarbrücken
www.vdm-vsg.de

Telefon +49 681 3720 174
Fax +49 681 3720 1749

Die VDM Verlagsservicegesellschaft mbH vertritt

Printed by Books on Demand GmbH, Norderstedt / Germany